Praise for *The Inter*

"No doubt you've heard, 'Timing is everything?!' I don't agree that it's everything, but timing is often the difference between success and failure. If there was ever the perfect time for *The Intentional Engineer* to be published, it is now! The aftereffects of the COVID-19 pandemic (layoffs and changes in where you work) resulted in thousands of engineers questioning their futures. Jeff Perry draws on his personal experiences and those of people he's helped become intentional engineers. Regardless of where you are in your career, I recommend you read *The Intentional Engineer* and become more intentional in shaping your future."
— John A. White, Jr., Chancellor Emeritus, University of Arkansas, Distinguished Engineering Professor (retired)

"In *The Intentional Engineer*, Jeff Perry masterfully guides engineers and technical professionals through a transformative journey towards a purpose-driven career. Perry's insights challenge the conventional norms and urge readers to design their lives with intention, aligning their unique genius with their career aspirations. This book is not just a manual; it's a call to action for every engineer to rise, introspect, and pave a path filled with intention and clarity."
— Hüseyin Kilic, Founder & CEO of Interesting Engineering

"Intentional living is the art of making our own choices before others' choices make us. Jeff Perry has written a book that teaches engineers how to create a beautiful life on purpose—using the mindset and skill set of innovative engineers. Must read."
— Richie Norton, Bestselling Author of *Anti-Time Management*

"To be successful in today's world, an engineer's mindset is just as important, if not more, than one's skill sets. In this book, Jeff Perry provides practical insights for building the right mindset to win in one's career and life."
— Anthony Fasano, PE, Bestselling Author of *Engineer Your Own Success*

"Jeff Perry 'gets it!' And, you can 'get it' too! The key to success is intentionality. And, the key to life-elevating intentionality is developing the right mindsets. In *The Intentional Engineer*, you'll learn how you can upgrade yourself so that your life and career take off."
— Ryan Gottfredson, *Wall Street Journal* and *USA Today* Bestselling Author of *Success Mindsets* and *The Elevated Leader*

"If you want to just hope that everything in your engineering career will turn out how you desire, then go ahead and put this book down. It's not for you. However, if you have goals of where you want to end up and actually be in control of your engineering journey, then this book is for you. Jeff does a great job of pulling out the key decisions you need to make along the

way and then gives practical and actionable guidance on how to make those decisions."
— Sol Rosenbaum, PE, Founder of The Engineering Mentor (www.TheEngineeringMentor.com)

"I chose to study Engineering at University because I didn't want to talk to people. I then realized that without communicating, I'm not really an engineer. I then noticed that if I don't take control of my career, my career will happen to me and I will end up anywhere, except where I want. That's why I'm glad to see this book covering these two topics (and much more) so folks like me can get control of their career, which also leads to getting control of their lives, and moving forward with intention."
— Ramzi Marjaba, Founder of We The Sales Engineers

"This is a great book to help those who want to change their course in life. It is a roadmap for understanding where you are and where you want to be, and it provides the tools to help you get there. I will implement what I have learned in this book and am confident these changes will lead to a more rewarding and successful career!"
— Sam LaMontanaro, PE, CEM, Director of Engineering at Aufgang Architects

"Jeff Perry's much-needed focus on the impact of being intentional in our engineering roles is a breath of fresh air to an ever-more entitled workforce. *The Intentional Engineer* is an essential read for anyone that wants to see beyond the daily grind of colorless work and truly make a difference in the lives around them. The outcomes to those around us and to our business will be important. But the change to us will be truly transformative."
— Colin S. Campbell, President & CEO and former VP of R&D, METER Group, Inc.

"I highly recommend *The Intentional Engineer* to anyone navigating the dynamic landscape of the tech sector. In a time of rapid change and uncertainty, this book provides valuable insights into aligning your career with your personal values and purpose. It's a refreshing perspective that goes beyond technical skills, emphasizing our work's profound impact on our professional and personal lives."
— Sandeep Seshadri, Executive Vice President of Engineering at Kasasa

"An engaging, informative book that offers clear explanations and step-by-step guidelines for identifying and reaching your values and goals, both personally and professionally. A tremendous resource for engineers, scientists, and other technical students and professionals who are seeking to build a happier, healthier, more intentional life."
— Katy Luchini Colbry, PhD, Assistant Dean, Engineering Graduate Student Services, Michigan State University; Director of Engineering Futures professional development program of Tau Beta Pi

"As engineers we spend so much time designing, managing, and analyzing the world around us. But do we take the time to design our own career and life? Let's face it, we can all take a fresh look at our life as a whole, and this book provides the inspiration and tools to do just that."
— Andy Richardson PE, SE, Principal Engineer at 29E6, LLC

"In this timely book, Jeff Perry shares a wealth of practical tools and personal stories to help technical people realize they have many more options and opportunities for professional and personal satisfaction than they could imagine."
— Leo MacLeod, Author of *Coaching and Mentoring For Dummies*

"In *The Intentional Engineer*, Jeff Perry has done an outstanding job of guiding the engineering professional through many of the challenges faced by most of us as we have navigated careers in engineering. It's a great guide for mentors who interact with students and young professionals. The philosophy of the book is one of guiding your own destiny and acting instead of reacting. It is an enabling philosophy for anyone but is particularly important for young engineering professionals who often tend to focus on details of the present task and do not act in a way that dictates the direction of their career. This book would be an ideal gift for a graduating engineer who is about to embark on a challenging career."
— David Field, Director, Institute of Materials Research at Washington State University

"Jeff Perry's book is a game-changer for anyone looking to transform their career into a meaningful and satisfying journey. Whether you're a student or a seasoned professional, it guides you through a process of self-reflection and life design. In a world where job dissatisfaction is prevalent, Jeff's book empowers individuals to create a path that brings a sense of purpose and fulfillment."
— Monte Marshall, Assistant Dean, BYU College of Physical and Mathematical Sciences

"It's been said that a life without a plan is like a ship without a rudder. That rudder is our intention, and without it we're doomed to drift aimlessly through the oceans of our lives and careers. In *The Intentional Engineer*, Jeff masterfully explains how to set and use intention to achieve the satisfaction and joy we all crave from our work. Jeff lays out a clear path with well-defined steps that guides readers through his proven process easily and efficiently. If you are an engineering professional who feels compelled to grow into a more fulfilling career, this book is a must read."
— Aaron Moncur, Founder, Pipeline Design & Engineering and The Wave: A Community for Engineers

"This is the mindset book every engineer needs, with practical, no-frills information and exercises on how to get clear on what you want, unpack negative beliefs that may be holding you back, and turn that self-awareness into intentional action. If you're an engineer or technologist who has found

yourself on autopilot in your career and unhappy with your current path, this book is for you."
— Stephanie Slocum, PE, Author of *She Engineers* and founder of Engineers Rising, LLC

"Jeff has hit on something absolutely critical with his book. In this *Age of Distraction*, it's all too easy to get swept up in the busyness of our work and lose sight of what really matters. As a result, living and working with intentionality has never been more important. If you don't choose your own intentions, someone else will gladly do it for you. Follow Jeff's wit and wisdom closely, and you'll put your career—and your life—on the path to success."
— Patrick Sweet, P.Eng., MBA, CSEP, PMP; President of The Engineering & Leadership Project

"We are facing a crisis of awareness and intentionality despite the availability of information and technological advancements. This crisis affects engineers as much as everyone else. Jeff's timely book reminds us of the importance of understanding our values, developing self-awareness, and becoming more intentional in our actions. Only when we live intentionally can we move towards a more fulfilling life and career."
— Henry Suryawirawan, Host of Tech Lead Journal Podcast

THE
INTENTIONAL
ENGINEER

THE
INTENTIONAL
ENGINEER

A GUIDE TO A PURPOSE-DRIVEN
LIFE AND CAREER FOR ENGINEERS
AND TECHNICAL PROFESSIONALS

JEFF PERRY

PALOUSE
PUBLISHING

Book design by Brynn Steimle
Cover design by Robin Perry

First edition 2023.
www.jeff-perry.com
Published by Palouse Publishing.

This book is for Robin and our four beautiful children. Living a joyful life with you is a fabulous intention.

CHECK OUT THE INTENTIONAL ENGINEER WORKBOOK!

The Intentional Engineer Workbook is a companion book to this book. I have created it to assist you in completing the activities and exercises provided throughout this book.

Just to say "thank you" for reading my book, the digital version of the workbook is **free**—it's my gift to you. You can download it at **www.TheIntentionalEngineer.com/workbook**. You can either print the digital version or type your answers right into the PDF.

Or, if you prefer to write your responses by hand, you can purchase the workbook on Amazon.

Enjoy!

Jeff Perry

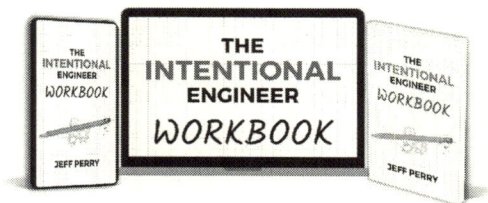

TABLE OF CONTENTS

THE
INTENTIONAL
ENGINEER

INTRODUCTION

YOUR CALL TO BE INTENTIONAL

Intentional living is the art of making our own choices before others' choices make us. —Richie Norton[1]

I WAS LOST...

In the summer of 2019, I was completely lost when it came to my career. I had *no* idea what I wanted to do next.

I was just a few years out of my undergraduate degree in mechanical engineering, and I had done well for myself up to that point—I had moved to a new company with a significant pay increase, was in a leadership position, had spent time building side hustles selling products on Amazon, had a lovely wife and two amazing kids, was halfway through an MBA degree at a prestigious university, and we owned our first home.

But I was still lost.

My role at work had given me a lot of opportunities to progress and grow. I had built relationships with new colleagues, expanded my leadership capabilities, facilitated company-wide training and coaching sessions on mindsets, and been trusted to take care of high-profile cross-functional initiatives.

But I was still lost.

If you had asked me ten years earlier what success would look like at this point in my life and career, I was checking all the boxes. So this whole feeling lost thing came with confusion and a bit of guilt. Everything looked the way it "should"...so what was wrong? Something inside me said I was headed in the wrong direction.

Here are some of the circumstances that made me feel this way:

- As part of our company's restructuring, I had moved from being a unit director reporting to the executive team to reporting to another manager.

- Many of my large projects were well underway, but since I had already created the necessary systems, I was now maintaining these programs and systems rather than creating new initiatives. So I was no longer doing the work I enjoy, work that makes me thrive.

- The mindset training and coaching I had started doing was some of my favorite work I had ever done, but it was such a small part of my daily work—probably less than 10%.

Because of these factors and others, I knew I needed to make a change, but I didn't know what change I needed to make.

I was lost.

The process of finding my desired path and making it a reality was tough and continues to evolve more than four years later. But along the way, I've learned a lot that has helped me and allowed me to help thousands of engineers and technical professionals through coaching, training, webinars, podcasts, speaking, writing, and more. It's been amazing to see transformations such as:

- Kim, a PhD engineer who changed from working in a small consultancy to running technical projects at a national lab, initially increasing her earnings by 40% and then receiving three raises in less than two years!
- Nitin, who has a master's degree, who found himself unemployed for months, but ended up with three job offers and a promotion in his first year working with me!
- Jennie, who left a toxic work environment and was able to move to a great organization, working in a desired role that brings together her engineering prowess with her passion for sustainability.
- Henry, who was searching for his "happy place" in his career, and realized he actually needed to change himself. Once his mindset shifted, the external factors fell into place and he was able to move into a new job where he was supported in all areas of his life and received desired growth opportunities.

The number one principle all of these people applied to make these changes is this: *be intentional*.

By living with intentionality in my own life and helping others to do the same, I have created my life's work by helping others create theirs!

THE CURRENT STATE OF CAREERS—IS THIS IT?

The data on career satisfaction these days doesn't always paint a pretty picture. Many people seem to be unhappy and unengaged in their work. For something that consumes 40+ hours per week for most people, that isn't a recipe for a society overflowing with joy.

For example, Gallup's State of the Global Workplace: 2023 Report shares that:

- Only 23% of employees feel like they are thriving at work, while 59% are "quiet quitting" and 18% are actively disengaged.
- Fifty-one percent of employees are watching for or actively seeking a new job.
- Worldwide, 44% of employees experience significant stress on an average day (52% in the USA).[2]

Interestingly, this data seems to clash with some other data, particularly that from a survey conducted by the Conference Board. Their 2023 report on job satisfaction shares:

- Job satisfaction is at the highest level since the survey began in 1987, with overall satisfaction at 62.3%.
- Women are significantly less satisfied than men across most satisfaction components, which means there is a lot of work to do to address the gender gap.
- The top factors that separate those who intend to stay at a company vs. those who intend to leave include wages, organizational culture, leadership quality, and work-life balance.[3]

Even if we assume the higher rate is correct and more than 60% of employees are satisfied with their jobs, that leaves millions of people who are *not* satisfied.

At some point (usually multiple times in our lives), we start asking ourselves questions like, "Is this it?" "Shouldn't there be more to my career and life?" and "How do I figure out what I want to do and make my goals and dreams a reality?"

I have definitely felt, at times, like I wanted more out of life, but after hundreds of one-on-one conversations with other engineers, I know it's not unique to me.

The message of this book is this: Yes, you can have more! You don't have to be lost! You can create intentions that matter to you, and turn those intentions into your living reality. It's not easy, but it's 100% worth it!

THE RESPONSIBILITY OF ENGINEERS

Engineering is the art of directing the great sources of power in nature for the use and convenience of man. —Thomas Tredgold[4]

Engineers and tech professionals are world changers. The world is rapidly evolving because of innovative technologies and engineering solutions. As technology continues to advance, it is increasingly important to consider the social responsibilities of engineers and technology professionals.

The word "engineer" encompasses so many different duties: designing and building infrastructure, developing technologies, and creating products we use every day. The work we do has a significant impact on society, and we shouldn't take this responsibility lightly.

There are a variety of viewpoints on the subject of the social impact of engineers. Some argue that engineers have a duty to prioritize the safety and well-being of society above profits and personal gain. Others believe that engineers should focus solely on designing and building safe and efficient products and infrastructure, leaving social responsibility to other professions. Regardless of individual opinions, engineers and technology professionals have a responsibility to at least consider the impact their work has on society and the environment.

Professional Engineers (members of the National Society of Professional Engineers) in particular have a Code of Ethics and a Creed that serve as a guide. The Engineers' Creed states:

As a Professional Engineer, I dedicate my professional knowledge to the advancement and betterment of public health, safety, and welfare.

I pledge:

- *To give the utmost of performance;*

- *To participate in none but honest enterprise;*

- *To live and work according to the highest standards of professional conduct;*

- *To place service before profit, the honor and standing of my profession before personal advantage, and the public welfare above all other considerations.*

In humility, I make this pledge.[5]

Whether you are a Professional Engineer or not, that's a good creed to live by. Take pride in your work and do your best, because what you do impacts the lives of others!

THE OPPORTUNITY FOR YOU

Time is not refundable. Use it with intention. —Unknown

In addition to a great *responsibility*, there is also an amazing *opportunity* for engineers to do work that advances our lives in many ways! The technology you create, the infrastructure you build, the products you design—they define small and large parts of the lives of people around the world.

Here are just a few possibilities:

- You might be ensuring the safety of the products you are working on to save lives.

- You might be standing up for positive digital engagement and oversight for young people to promote improved mental health and well-being.

- You might be increasing efficiency or sustainability in meaningful ways to be a wise steward of our natural resources.

- And so much more!

The potential you have to create change is a great responsibility, and something you must take seriously. If you don't, the consequences could be devastating. This is true no matter what your current occupation is. It's about your potential. So if, currently, you feel like your work provides little meaning to yourself or others, there are opportunities all around you that will help you realize your full potential.

What you do matters, and your work will take up a large part of your life—why not be intentional about it and take responsibility to use your skills for good?

CONSIDER YOURSELF "CALLED"

An object at rest stays at rest...unless acted upon by an outside force.
—Newton's first law of motion

Do you feel like you are "at rest" in some way in your life and career? Maybe you feel stuck in a role you don't enjoy, pigeon-holed into a career direction you don't want to take, or like work is taking over your life and you have no time to spend doing things you love with people you love.

If that's the case, let this moment be a turning point for you. The moment when you decide to be intentional, and invite an "outside force" to shift you from rest to movement.

Take time to answer questions for yourself such as:

- What do you want to be when you grow up?
- What impact do you want to leave on the world?
- What do you want others to remember you for?
- What characteristics or skills do you want to grow and improve?
- What kind of person do you want to become?

- What is the next right action you can take?

Rather than frustrate you, let these questions excite you! No one else gets to decide how you answer these questions. While your life circumstances in the past might make it harder or easier compared to others to realize your dreams, you still get to choose how you act given your current reality.

Have you read *The Hobbit*? J.R.R. Tolkien writes about Bilbo Baggins, a character who loves the comforts of home, but also feels a pull toward something more. Then some "unexpected guests" show up and invite him on an adventure (where there will be dwarves, wizards, dragons, elves, trolls, and much more). In reply to their invitation, Bilbo says, "Sorry! I don't want any adventures, thank you. Not Today. Good morning! But please come to tea— any time you like! Why not tomorrow? Good bye!"[6]

Perhaps you, like Bilbo, feel you're being pulled toward something new and unknown, but keep holding back. If you have read *The Hobbit* or watched the movies, you know that Bilbo eventually chooses to join the quest, even though he had no idea what was in store for him.

The story of Bilbo is a great example of what Joseph Campbell referred to as the "Hero's Journey," which is a common structure to many stories, myths, movies, etc.[7] Bilbo is in what Campbell called the "Ordinary World" (the comforts of the Hobbit hole) and then gets a "Call to Adventure." While he initially refuses this call, he encounters a mentor (Gandalf the wizard) and eventually crosses the threshold and moves into action, leading to experiences that change and transform him.

I've realized that one of the reasons I was discontent with my life years ago was that I was following someone else's path. I felt called to more, and in stepping out into the unknown, I've embarked on my own adventure. I've found mentors and guides, and I've been a guide to others. I'm still learning, as we all are, but

I'm committed to living a life of intention rather than comfort, as that is the path of transformation.

As you read this, this moment can be the beginning of *your* Hero's Journey! You have a decision to make. You have the opportunity to create your life and career with intention—will you take it? Don't stay at rest and expect things to change on their own. You're being called to do and be more—answer it!

BECOME AN INTENTIONAL ENGINEER

Our intention creates our reality. —Wayne Dyer[8]

As you can imagine, when Bilbo Baggins returns to his home in the Shire after his adventures in *The Hobbit*, he is a very different hobbit than he was when he started. He went through challenges, loss, pain, and battles, and experienced new depths of friendship and belonging. He gained greater confidence and resilience, as well as increased awareness of the world and his place in it.

The same can be true for you.

The process and principles in this book are about making intentional decisions, visualizing how to make those decisions a reality, and then committing to work towards your decisions each and every day. Focusing your intention each day, week, month, quarter, year, and for your life will move you toward who you want to be and help you live this one, precious life in a way that you are proud of.

Yet this book is not intended to help you arrive at one certain end point. If you read it and take the recommended actions, you will make progress towards meaningful goals. That's awesome! But more important than whether or not you achieve specific goals or not is the process of *becoming*. You want to *become* intentional, someone who lives their life with purpose, and takes action to create outcomes you desire. This process will transform you.

Abraham Lincoln said, "Every man is born an original, but sadly, most men die copies."[9] Don't let that be you! Be your own person! This process of changing and transforming takes time, but it's worth it.

TAKE INTENTIONAL ACTION

At the end of each chapter of this book, I will invite you to take specific actions to help you apply the concepts and principles from that chapter. Certainly you don't have to do 100% of the actions and activities, but take the ones that seem most helpful to you and put them into action. Otherwise, you'll simply gain ideas without anything changing. Here's your call to take intentional action on the principles from this chapter.

Keep a Journal

Nothing can move you forward in your own learning more than a journal.... A journal provides a safe place where you can discover, think, reflect, plan, and dream. —Brad Wilcox[10]

One of the most powerful tools you can use as you seek to create a vision of your future and what you want to become is a journal.

There are many effective styles of journaling. You can try free-writing, reflecting on the highs and lows of each day, or Q&A journaling where you write down questions you want to ponder (like many found in this book) and give yourself space to write the answers as they come to you. The format doesn't matter as much as the practice of writing regularly.

I write in my journal every morning and every night.

In the morning, soon after I wake up, I spend time writing about what I plan to accomplish that day, week, month, and year. I also write about who I am becoming. I write it in *present* tense.

Sometimes I write about goals. And sometimes I write affirmations in the form of "I am" statements that I am working to make true. As I write them, they become more true for me each day. I then spend some time experiencing the feelings of what it is like to *be* the person I am writing about.

I also use this time to brainstorm in my journal ideas for things I want to write or create, or ways to help my clients. Doing this in my journal rather than on my computer helps me avoid getting distracted.

Writing by hand is also much slower than the speed of thought, so it forces me to slow down a bit, allowing my mind to explore more ideas than just the ones I'm writing at the time. It's pretty cool.

At night, I review my day, write how it went, and what I could have improved. I think briefly about what I want to accomplish the next day, and record no more than three priorities I will work on.

Finally, I consider a question I want to ponder while I'm sleeping. Your subconscious is extremely powerful—your brain can solve a lot of issues while you're sleeping. As Thomas Edison said, "Never go to sleep without a request to your subconscious."[11]

Writing in a journal is extremely powerful. It's something I expect of all of my coaching clients. Start now if you aren't journaling already. And keep your journal handy when reading this book—this book frequently recommends journaling about the ideas I'm discussing.

Answer These Questions

To get started journaling, turn off other distractions and spend 10 minutes right now answering the questions in the "Consider Yourself Called" section, reprinted here:

- What do you want to be when you grow up?
- What impact do you want to leave on the world?
- What do you want others to remember you for?
- What characteristics or skills do you want to grow and improve?

- What kind of person do you want to become?
- What is the next right action you can take?

Your answers may evolve over time, and it's okay that what you write isn't perfectly polished. What's important is that by beginning the process of being intentional, you are setting your course to be more aware of opportunities around you and consciously define the actions you would like to take.

These and similar questions will help you uncover the insights you have been searching for.

Get out of your head and write in order to get clarity about what you want in your career. See what you discover.

To support your application of the book, get your FREE download of *The Intentional Engineer Workbook* at www.TheIntentionalEngineer.com/workbook.

CHAPTER 1

START WHERE YOU ARE

Do what you can, with what you've got, where you are.
—Squire Bill Widener[12]

WHAT IS THE CURRENT STATE OF YOUR CAREER?

Take a look at the model below, and consider the current state of your career and life:

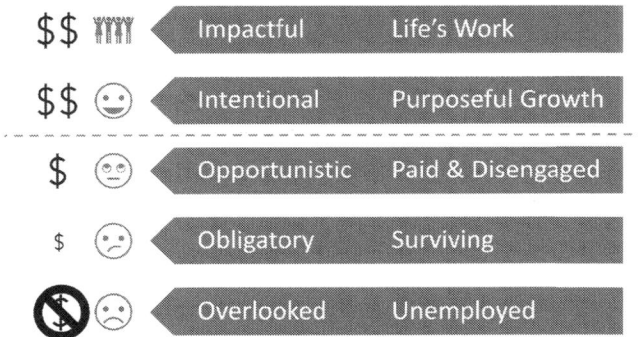

Figure 1.1: Intentional Engineer model

Which level do you find yourself at right now?

If you're unemployed and feeling overlooked, it's hard to get past that, so you may be at the bottom for now. That is a hard place to be.

If you have a job but it's barely making ends meet and you're in survival mode—that's better, but still not great. You're working in obligation, which isn't very fun.

Perhaps you're in a cushy job, making plenty of money as opportunities come your way. Yet on further reflection, you find yourself in a situation that others laid out for you, not one you really want to be in. So you're paid well but disengaged from your work.

This is perhaps the most dangerous place to be, as it's a place where many settle and get comfortable because it's "good enough." You may be making money, have some happiness in your family, and have moments of satisfaction. Yet, when you start your day, is it with joy for what awaits? Are you truly excited about the challenges you will face? Do you experience purposeful growth in your life—do you keep improving and adding new and novel dimensions to your life? Is your life already bigger than you previously imagined possible?

If not, that's okay for now, but the tipping point and the factor that changes everything is being intentional.

Being intentional in your life and career means *you* are defining what is important to you. When you're intentional, you align your actions, work, and focus to move towards what you want and who you want to become. That's why I call it "purposeful growth." You're not stagnant, but are growing in a direction that is important to you.

You get to choose. Exciting, right?

Shifting towards living intentionally will facilitate the type of growth that, over the course of your life, will allow you to have an impact on others because of who you are and the work that you do.

You can lead and mentor others, you can create products and services that change lives, and you can be a positive influence in your family, your community, and the world, as you contribute to causes you care about. You don't just enjoy your life and career for yourself, but what you do impacts and blesses the lives of others as well.

So no matter where you find yourself on the Intentional Engineer Model today, you can choose to move up. You can't start from anywhere other than where you are right now. Accept your current reality, but start creating your future with intention!

INTENTION BEGINS CREATION

> *Be mindful of intention. Intention is the seed that creates our future.* —Jack Kornfield[13]

Intention is the beginning of creation. We can't accomplish anything without first starting with a thought or an idea.

As Stephen R. Covey, author of the classic leadership book, *The 7 Habits of Highly Effective People*, said:

> *All things are created twice. There is a mental or first creation, and a physical or second creation.... If we do not develop our own self-awareness and become responsible for first creations, we empower other people and circumstances...to shape much of our lives by default.*[14]

The act of intention is an act of decision. You get to decide the life and career you want to create. You decide the kind of person you want to become.

This intention pulls you toward the impactful life you want to live.

Living intentionally invites us to have an *approach-orientation*, which means you are growth-oriented. You're willing to take appropriate risks for the opportunity, not the guarantee, of a better future.

Conversely, you could have an *avoid-orientation*, which means you fear risks and fear failure. You don't want to lose, and you mostly focus on maintaining what you have now.[15]

Which approach do you think is likely to lead to the results you want?

That's right, approach-orientation!

Your goals and focus in life will be much more powerful if you are focused on what you are moving *towards* rather than what you are trying to move *away from* or *avoid*, because as Tony Robbins says, "Where focus goes, energy flows!"[16]

This comes to life in many career-development conversations I have with people who express the need for help. When I ask what they want in their careers, they often tell me what they *don't want* or what they want to get away from. They tell me things like:

- "I don't want to feel so stuck with what I'm doing now."
- "I want to get out of this bureaucracy."
- "I need to get away from my narcissistic manager."
- "My current company is imploding, and I need to get out of here."
- "I'm unemployed right now and not getting any interviews, that needs to change!"
- "I want to stop getting passed over for promotions."
- And on and on....

Instead of these avoid-oriented goals, we work together to focus on what they really want, and create approach-oriented intentions like:

- "I want to work in a role that allows me to unleash my creativity and innovative ideas."
- "I want to experience growth in my skills as well as my income by X amount."

- "I'd love to work for a supportive leader who promotes collaboration and psychological safety."
- "I want to work for a company that is stable and growing so that I can feel secure in my role there."
- "I want to improve my job search approach, interviewing skills, and more to find a new opportunity that leverages my experiences and passions."
- "I want to present my value as an employee in an effective way so that I'm more likely to be promoted."
- "I want to increase the positive influence I have on my team."

Do you see the difference in the energy behind these statements?

Focus on what you want and you're more likely to find a way to get it!

EMBRACE UNCERTAINTY

Most people prefer the certainty of misery to the misery of uncertainty. —Virginia Satir[17]

One of the challenges of living a life of intention is embracing uncertainty. As Satir's quote suggests, sometimes it feels safer to just stick with what we know rather than make a change, even if it is miserable!

But is that any way to live? I don't think so. And I don't think you do either, which is why you're reading this book!

Here's another quote I love, from Brené Brown's book, *Atlas of the Heart*:

Choosing to be curious is choosing to be vulnerable because it requires us to surrender to uncertainty. We have to ask questions, admit to not knowing, risk being told that we shouldn't be asking, and, sometimes, make discoveries that lead to discomfort.[18]

Whether we like it or not, dealing with uncertainty is a fact of life. We never know exactly how things are going to play out.

Don't let fear of uncertainty lead you to inaction. Take the information you have, make your best judgment, and move forward. If it doesn't work out, that's fine, you can try again!

Your life and career is a series of experiments, but if you stop running experiments because you are afraid of the potential outcomes, you won't collect much data! So keep experimenting.

My client Hector is a great example of someone who didn't let fear of uncertainty stand in the way of making the career change he knew he needed. He was a mechanical engineer with excellent experience in large and small companies. After a few years with a startup, he realized the company culture was holding him back and micromanagement got in the way of his growth and ability to apply his skills. He knew he couldn't thrive if he stayed where he was at, so he made the difficult choice to quit.

This decision brought on all sorts of uncertainty. He had a family to take care of and no known prospects for a new opportunity. (It's important to note that I generally wouldn't recommend doing it this way, but it was what he felt strongly he needed to do.)

It was shortly after he left his company that I met Hector and we started working together. He needed help charting his path to what was next. He needed to land a new opportunity, but also had desires to start his own business creating products he was excited about.

Luckily, I knew an engineering services firm owner who needed someone with Hector's skills. I connected them, and within three days, Hector had landed a contract job. A job that would allow him to work remotely while working on his new business and looking for a more permanent, full-time role, which he found in just a few short months using principles we worked on together.

Now he's thriving in his role and working towards increased leadership opportunities to grow his influence while working on an Executive MBA degree!

But nothing changed until he stepped out into the unknown.

EXPECT RESISTANCE

In his book, *The War of Art*, Steven Pressfield states:

> *Remember our rule of thumb: The more scared we are of a work or calling, the more sure we can be that we have to do it.... Resistance is experienced as fear; the degree of fear equates to the strength of Resistance. Therefore the more fear we feel about a specific enterprise, the more certain we can be that that enterprise is important to us and to the growth of our soul. That's why we feel so much Resistance. If it meant nothing to us, there'd be no Resistance.[19]*

Getting from where you are now to where you want to be will take a lot of effort. If changing is truly important to you, there will be resistance.

Resistance comes in many forms. Here are just a few examples of resistance I have seen in my clients and experienced myself:

- Allowing fear of many things such as failure, what others will think, or of the unknown to inhibit action.
- Turning to distractions that are easier and simpler to engage in than the hard work of change, such as games, social media, news, TV/movies, etc.
- Being slowed down by impostor syndrome with thoughts like, "Why me? What if they find out my limitations and weaknesses?" (Yes, I'm feeling plenty of that even as I write these words.)
- Holding back because of concerns about market conditions, layoffs, or other external factors.

As you consider the changes you want to make, what resistance are you feeling now? What is pulling you back to comfort and a perception of safety?

If you can identify it, you are more likely to be able to push through it!

You can expect resistance, but don't let it be your master.

LIFE-WORK ALIGNMENT

Do you work to live, or live to work?

Many of us believe that we place a higher value on faith, family, health, relationships, and hobbies than on work responsibilities, but our attention and priorities don't align with that. It feels like work is in opposition to the life we want to live rather than enabling it.

As Parkinson's law states, "Work will expand to fill the time allotted for its completion."[20] So if you allot most of your time to work, it will fill all of that time, and your life will only have the little space that remains.

The emails keep coming, the project deadlines are looming, and there are so many expectations that we give all our time and energy to work. Then all the other areas of our lives have to squeeze into whatever time is left in the day after work. Sometimes it feels like we don't have time or energy for anything more than getting takeout food and watching TV day after day. I'm not saying work should *never* be the priority—perhaps there will be periods of your life and career where your number one priority is taking care of big work responsibilities, but it's typically unsustainable to work that way for an extended period of time.

Achieving *work-life balance* is often the prescribed solution. But notice—when the phrase is worded in this way, "work-life balance," work comes first. If we are truly going to make living an intentional life our priority, we need to seek something different. We need to swap the priority levels of work and life, creating *life-work balance*, where life comes *before* work. With life-work balance, life expands to fill the space we give it and work fills the space that remains.

Now, you may be saying to yourself, "But work already takes up so much of my life and I'm so busy. How am I supposed to give *less* time to work when I already feel overwhelmed there?"

Well, there is a high likelihood that much of your time at work is spent doing tasks that are not only unimportant and unnecessary but also not impactful. This could include creating reports that rarely get used, sitting in meetings going over the same things again and again, and, let's be honest, probably spending time aimlessly browsing the internet or checking your email.

With a little intention, we can make our time at work much more efficient. The Pareto Principle, aka the 80/20 rule, suggests that 80% of our outcomes result from only 20% of our efforts.[21] If you shift your focus toward that 20%, and even do more of those impactful tasks rather than getting lost in the 80%, you can actually accomplish *more* in *less time*, and the demands of work actually begin to focus our efforts rather than get us lost in them.

Less time spent on less important tasks means more time to be balanced on the life side of the balance equation. Life can become the priority.

If you start to make changes in this direction, you might find that you like what you experience. Life-work balance is better than work-life balance, but is it really *balance* we are after?

Probably not. Think about it, if things are in balance, there is no motion. It means stagnation. It is equilibrium in the perceived tug of war between life and work.

But as an Intentional Engineer you know it's not just about more recognition or making more money, you have a calling to satisfy the desire to do more and be more. You aren't satisfied with stagnation.

So rather than balance, what you are really after is: *life-work alignment.*

Alignment allows your life and work to be in sync with your overall values and mission, working together for a common purpose. Work isn't in opposition to life, it supports it. Life also complements your ability to do great work.

It's like the addition of vectors in mathematics. They have a magnitude and a direction. If you're trying to balance your life and work, it feels like they are in opposition to each other and get to zero, like this:

Life ⟶ ⟵ **Work**

Figure 1.2: Life-work balance: work and life perfectly balanced

But if your life and work are in alignment, they will add together to make something even greater, like this:

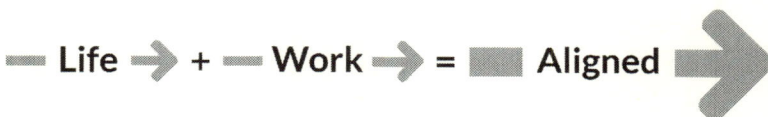

Life ⟶ + **Work** ⟶ = **Aligned** ⟹

Figure 1.3: Life-work alignment

Your life and work don't always have to work against each other. In fact, if you integrate and align them, you're likely to have greater results.

For example, recently I decided to sign up and train for a triathlon. I had been running occasionally and engaging in other workouts, but not swimming or biking at all. Since the bike course is the longest leg of the race, I really needed to spend time on my bike.

Along came a Saturday that my family wanted to go to our local farmer's market. There was a bike shop at the market, and my bike needed a tune up. Also, I was aware that one of my clients from out of town who I wanted to catch up with would be attending the farmer's market.

So I aligned the day. While my family drove over, I rode my bike to the farmer's market to get some training in. I took my bike to the shop. I got to catch up with a client and bumped into a friend too, all while enjoying a fun time with my family and buying delicious fresh fruit from local growers. What a day!

Not every day and task can integrate so well, but if you get creative, you'll likely find more ways to intersect and align your life and work than you currently anticipate.

TAKE INTENTIONAL ACTION

Now it's time to start creating your intentions by journaling. Remember, what you write doesn't have to be perfect. Journaling is a great way to figure out who you want to become.

Reflect on Where You Are

Spend 10 to 15 minutes in your journal writing and reflecting on your current situation. What are the things you like and don't like about your life and career right now? Get it all out.

You may also want to review the Intentional Engineer Model (figure 1.1) at the beginning of the chapter and get clear on what level you're on now, and where you've been in the past.

While you may not be happy or content with the current state of your career, instead of just focusing on what you want to change, look at the progress and achievements you've enjoyed in any area of your life over the last three to five years. Make a simple list. You could include (but are certainly not limited to):

- Amazing friends or mentors you've met
- A new job or promotion you got
- A degree or certification you completed
- Getting married or having children

- Any improvements you've made to your health
- New skills you've learned or improved on
- Travel experiences that have expanded your horizons
- Improvements to your spiritual life
- Moving to a new city or home
- Being able to support someone else who needed it
- Or *anything* else positive that you want to celebrate!

The purpose of going through this exercise is to realize that even if you haven't been fully intentional, you've changed and made progress in the last few years.

Now imagine how much *more* you will change and grow over the next few years as you live with increased intention!

List Expected Resistance

As was stated in this chapter, you can expect resistance to come your way as you work towards living a life of purpose and meaning. Naming that resistance and how it usually impacts you can help you move through it.

Spend five minutes making a list of how resistance usually manifests itself in your life. What kind of resistance do you expect as you begin living intentionally? (See the "Expect Resistance" section earlier in the chapter for some ideas.)

Bonus—for each resistance you came up with, come up with at least one way you can counteract it.

Make a List of What You Want

Now that you have looked back on your progress and have identified your likely areas of resistance, it's time to start thinking about what you really want in your life and career.

Again, this doesn't have to be perfect. Don't get caught up in the details. Just list things that excite you. You can reduce and edit it later. For now, just get it out. Here are some ideas to get you started:

- What kind of family do you want to have?
- Where would you like to live?
- What kind of car do you want to drive (if any)?
- What do you want to learn?
- What experiences would you like to have?
- How much money do you want to make?
- How do you want to help others?
- Are there causes or charities you want to contribute to?
- What kind of spiritual life do you want to have?
- What is your ideal level of health? Are there health-related experiences (like running races, etc.) you want to have?
- What hobbies do you want to make time for?
- What ways do you want to contribute to your community?
- How do you want others to describe you as a person?

You certainly don't need to answer every one of these questions now, and you may come up with others; the key is thinking about what kind of person you want to become.

CHAPTER 2

DESIGN WITH INTENTION

DESIGNING YOUR LIFE AND WORK

When I first talked with Mark, he was an experienced engineer who had spent about 10 years in the defense industry working on things like Naval warships and advanced laser weapons for the US Military. He had also changed from being a mechanical and structural engineer into being a systems engineer.

But he was at a crossroads in his career. He wanted a change—he had lost some of his passion for the defense industry, and saw some of his friends making significantly more money and working on exciting projects. Having lived in Seattle for the previous two years or so, he wanted to try his hand in the tech industry.

It was time for Mark to design the next phase of his life and career.

In their excellent book, *Designing Your Life*, Bill Burnett and Dave Evans use principles of design thinking typically applied to creating products and services to help us figure out how we can design our lives and careers. One of the key lessons from the book is this: Our lives are a series of experiments or prototypes. We get to decide something we want to do or try. We don't know exactly how it will turn out, but we need to go through the process of trying something different to see if it actually fits what we want.[22]

Frequently we look at people as "finished products," but know little of what it took for them to get where they're at.

Compare this with a great invention you think is amazing. Usually, you just see the final product. You don't see all the trial and error, the detailed research, and the many iterations it took to get there. The designers and engineers had to prototype until they got the product to the place they wanted it to be in order for it to be ready to bring it to the market.

We can do the same thing as we design our lives. We can create prototypes of ideas of who we want to be. Prototyping reduces risk as we explore our ideas. Good prototypes are inexpensive, quick to test, and easy to implement.

Rhianna Taniguchi shares two ways to prototype a new phase of life or career you are considering:

Prototype Conversations

You can have a casual chat or a more formal informational interview to learn from people who are doing things that you find potentially interesting and might want to design into your life. (Informational interviews are conversations with people in career roles you might be interested in, where you have an opportunity to interview them to gain information. Informational interviews provide the opportunity to give, be curious, learn, and build a relationship, and can last anywhere from 15 minutes to an hour or more.)

The purpose of prototype conversations is to learn from the experience of others, allowing you to make more informed decisions.

Don't limit your prototype conversations to only be with people who have careers you are interested in. You can have these conversations with people you admire who have great marriages or families, excellent health, solid spiritual lives, amazing friendships, or anything else you want to incorporate into your own life.

As you get ready for a prototype conversation, make sure to prepare good questions to ask in order to make the most of that time.

Prototype Experiences

Prototype experiences can take the form of volunteer work, an internship, job shadowing, or a side project as a freelancer or at your existing organization. Trying something on a smaller scale or short-term basis can help you decide if it's something you want to invest more time into.[23]

Prototyping ideas for your life means taking small steps that usually don't cost a lot to implement, so the risks are limited. If the prototype isn't successful, great! You learned something you didn't know before, and that can help you as you continue forward.

Eventually, if you are after a career shift or upgrade, you'll have to make the leap into that new role. Even so, you can still consider it a prototype, just a larger one. If it doesn't work out, there will always be an opportunity to make another shift later. But if you never make any changes, how will you know?

In Mark's case, he had multiple conversations with people in different companies and roles, and identified that a Technical Program Manager was very similar to what he was used to doing as a Systems Engineer. He validated and explored this option thoroughly and decided he wanted to try it out for real. In the end, he landed a job at a well-known company, nearly doubled his salary, and has enjoyed the change!

THE POWER OF CORE VALUES

It's not hard to make decisions, once you know what your values are. —Roy E. Disney[24]

One powerful way to live with intentionality is by identifying a set of core values—the beliefs and principles you want to live by.

Whether you realize it or not, you make hundreds and maybe thousands of decisions every day. When to get up, what to wear, how you talk to your partner/spouse, how you interact with your children, what you eat, what you write in that email—decisions happen moment by moment. Each decision you make is a reflection of your core values.

If you aren't intentional about identifying and living by your values, you might be surprised to see what your decisions actually express in terms of your values.

If, on the other hand, you base your decisions on your values, you are taking the opportunity to be deliberate by focusing on what is most important to you.

Identify Your Values

> *Your core values are the deeply held beliefs that authentically describe your soul.* —John C. Maxwell[25]

It may be obvious, but in order to live a values-based life, you must first decide what your values are.

When you identify your core values, they become a guide and a filter for all your decisions. Career, relationships, family, community—you have an opportunity to live each area of your life in alignment with your core values.

If we don't identify our values, our values get lost in all of our many distractions—our self-talk, to-do lists, and the day-to-day tasks we need to get done.

So write them down!

When you put pen to paper (or type them out), you make your values explicit, thus separating them from the thoughts in your head. You make them important in your life!

You'll then want to keep them somewhere you can see them frequently. Put the list on your bathroom mirror, in the front cover of your journal, near your desk, or anywhere else you will see them again and again. Writing them down and then forgetting them doesn't do you any good either!

Don't worry, I'll give you a process to identify your values at the end of the chapter.

Live in Congruence

Peace of mind comes when your life is in harmony with true principles and values and in no other way.
—Stephen R. Covey[26]

Once your values are identified, it's now time to live them. I love using the word "congruence" to describe living true to our values. You may remember from middle school geometry that congruence means that two geometric properties, like angle, side length, etc., are equal. One dictionary definition of congruence is, "The fact or condition of according or agreeing; accordance, correspondence, harmony."[27]

As stated in the quote by Dr. Covey, when you are in agreement, harmony, or living a life compatible with your values, you can have peace of mind.

Much of our distress comes when we are out of alignment with our values. If I state that I want to tell the truth but I lie to my wife about something, I'm out of congruence. This creates a tension internally that will grow and fester. I'm not at peace.

If I state that I want to have an impact on those I interact with, but interact with people primarily in ways to use them to accomplish my own ends, then my impact will be limited and I will be dissatisfied with my approach.

The opposite is true when you live in harmony with your values. Doing so brings growth, peace, and success in a sustainable manner.

Recommit When Necessary

The likelihood that you will live in congruence with your values 100% of the time with perfection is…low. Probably zero. And that's okay. All is not lost. When you find yourself living incongruently with your values, you can recommit and continue forward.

Identifying when you are out of alignment is a good thing! Doing so makes it possible to do something about it. Take a step back and consider what led to you going against your stated value. What happened? What was the situation? What were the thoughts or beliefs that led you to do it?

You can then make a plan to take further action to build your commitment to your values.

For example, if one of your core values is living a life of service, but in a moment of reflection you realize you rarely spend time helping other people, you might look for a volunteer activity to do in your community to help a cause you care about.

To use my previous example, if I lie to my wife about something, I need to first admit to myself that I'm not living in congruence. Then I can recommit and come back and tell the truth. That might be a difficult discussion to have, but it will get me back in alignment, strengthen my relationship with my wife, and give me peace of mind again.

We all stumble and fall at times—stay committed, identify adjustments and actions you can take, and keep working on living your values!

Sharing Your Values

One way to increase accountability and motivation to live your values is to share them with others! So I will share my values with you.

To help me remember them, I organized my primary values into an acronym—GIFTS (and I do believe that values, when lived by, can act as gifts in our lives).

- Growth
- Impact
- Faith
- Truth
- Significance

Each of these words has a certain definition that is important to me, and the words might mean something different to you. But that's the power of identifying and defining your own values—they have personal meaning, and that's what gives them power in your life.

MAKE DECISIONS

Do you ever get overwhelmed by the amount of options and opportunities available to you? It's amazing that we have so much freedom, but this wealth of options also presents its own challenges.

In his book, *Flow: The Psychology of Optimal Experience*, Mihaly Csikszentmihalyi said this:

> *The wealth of options we face today has extended personal freedom to an extent that would have been inconceivable even a hundred years ago. But the inevitable consequence of equally attractive choices is uncertainty of purpose; uncertainty, in turn, saps resolution, and lack of resolve ends up devaluing choice. Therefore freedom does not necessarily help develop meaning in life—on the contrary. If the rules of a game become too flexible, concentration flags, and it is more difficult to attain a flow experience. Commitment to a goal and to the rules it entails is much easier when the choices are few and clear.[28]*

I've seen countless people stuck in "analysis paralysis," mired in indecision because they were in between multiple good options. I've been there myself many times.

But we can't avoid decisions. In a very real way, to *not* decide *is* to decide.

It's interesting to look at the root of the word "decision." Part of the Latin root means literally "to cut off." The reality is, when we make a decision *for* something, we are also making a decision *against* something else.

This process can bring feelings of grief or loss as we emotionally let go of a possibility in favor of something else. That's okay. It also means we are taking greater responsibility because we are committing to something. That is scary, but it's a fear we must face.

Personally, sometimes I have avoided making decisions in the hopes that someone else would make them for me. Then, if it didn't work out, it wasn't my fault, right? If my wife and I were trying to figure out where to go for dinner, I wanted her to decide. If I was making a decision at work, I wanted to find a group consensus so that the decision couldn't be pointed back at me if things went wrong. If I was making a big life decision, I wanted to pray and ask God to tell me exactly what to do so I wasn't responsible for what happened as a result of my choice.

But abdicating responsibility is a feeble way to live. By being decisive and designing our own lives, we take responsibility and feel empowered.

So enjoy the freedom you have with all the options and opportunities available to you. Then, make decisions to narrow those choices down and chart your own path forward with intention.

TAKE INTENTIONAL ACTION

Identify Your Personal Core Values

Crafting a set of core values is a powerful process, and one that can have a large influence in your life.

Here is one way to go through the process.

1. List character traits or personal values you like in yourself and others—aim for 20 to 25 traits. To get started, think of words like fun, loving, forgiving, generous, optimistic, transparent, ambitious, spiritual, growth-oriented, etc.

2. Once you have a large list of character traits that you like in yourself and others, start to narrow it down to the traits you would most like to develop. You may have some that overlap or essentially mean the same thing. Group similar words together and pick the one you like the most, or just cross out traits that aren't as important to you. You should now have a list that is closer to 10 to 15 traits.

3. Now, narrow your list down to the essentials. Which of these traits are essential to how you operate in your life? These traits can become values for you to live by. Are there any traits or values that are good but that you don't want to be as large of a focus?

4. Make your final list—a list of three to seven values seems to be a sweet spot. Too few, and you don't cover much ground. Too many, and you're trying to be too much at once or won't be able to remember them.

5. Bonus—memorize your values. See if you can craft your values into an acronym or come up with another way to help you remember them!

Make a Commitment

There's a difference between interest and commitment. When you're interested in something, you do it only when it's convenient. When you're committed to something, you accept no excuses, only results. —Ken Blanchard[29]

Deciding to live a life of intention is not for the faint of heart. It's a process you'll need to commit to. While the process itself will be the quest of a lifetime, the decision to live with intention can be made in a single moment, right now.

This commitment is a commitment you're making to yourself. You don't necessarily need to share this with anyone else.

Below is a statement that you can sign and date as an indication of your commitment to living your life with intention. You can use the statement as is, or create your own. Either way, as you sign it, take a moment to feel the power of this decision and how it will guide your life moving forward.

I, _____ , commit to living my life with intention. I will live by my values, continually seek passion and meaning, and take empowered action to create a career and life that matters to me.

_____

Signature

_____

Date

CHAPTER 3

THE POWER OF MINDSETS

Once your mindset changes, everything on the outside will change along with it. —Steve Maraboli[30]

I'm not overstating it when I say that learning about mindsets has changed my life, and I wouldn't be doing what I'm doing or writing this book if I hadn't learned about it.

In 2017 the organization I was a part of was going through a merger with a smaller firm in Europe. As part of that merger, there was a decision to rebrand to a new, unified company with updated logos, company values, and more. That caused a lot of resistance in many employees, especially those who had been around a long time and wanted things to stay the same as they always were. There was even some significant contention at the executive and leadership ranks. The negative mindsets prevalent among many employees created a toxic culture at times and interfered with productivity.

Fortunately, someone on our executive team knew about the Arbinger Institute[31] and started engaging in discussions on how the principles that Arbinger teaches could help our organization.

Soon, one of Arbinger's consultants came to teach and train about 20 company leaders, myself included, in what they call the Outward Mindset. It opened my eyes to ways that I was being and acting with others in all areas of my life. I previously was unaware how certain ways I was interacting with coworkers, my wife, friends, family, and more were unhelpful, and in some cases, destructive. My eyes were opened.

After the training, I was so excited about the potential of the content and the changes its application could create in our organization that I volunteered to be one of the facilitators. I was able to deliver the training to about 200 people in our company over the next year or so. After I and other employees did the hard work necessary to shift our mindsets, the situation at our organization drastically improved.

Participating in the training sessions and seeing how the culture shifted in our organization helped me realize that what I enjoyed most about my work wasn't necessarily a technical innovation or completed project, but being connected to solving human problems and helping people, myself included, be better. That led me on the road to starting my coaching and training practice to help engineers shift their mindsets and live more intentional lives.

As I've engaged in this transformational work over the years, I've learned about and applied, and helped many clients apply, multiple mindset perspectives that lead to sustained changes. That's what this chapter is all about.

MINDSETS FUEL BEHAVIOR, BEHAVIOR FUELS RESULTS

Mindset drives and shapes all that we do—how we engage with others and how we behave in every moment and situation.
—The Arbinger Institute[32]

How is your mindset? Do you view other people as objects, or do you see the needs and wants of others as valid and important? Do you view yourself as a victim of your circumstances, or do you recognize your power to influence your circumstances? Are your default patterns of thought more positive or negative? Stop right now and answer those questions. Seriously, take 30 seconds.

Understanding and being aware of your mindset can unlock huge benefits for you if you take action and make changes in your life. This is because our mindsets are foundational to almost everything we do, but we often lack intention around curating the mindsets we want to have.

One of the problems is that when we want to make a change in our lives, frequently, we focus only on the actions and behaviors we need to change to reach our desired goals as opposed to recognizing that our fundamental way of operating (our mindset) needs to change. It goes something like this:

Step 1: Identify a new goal or set of results we want.

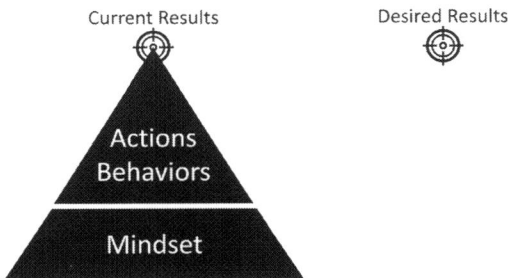

Figure 3.1: Mindset-driven results, adapted from the Arbinger Institute[33]

Step 2: Change our actions and behaviors to try and reach our new desired results.

Current Results   Desired Results

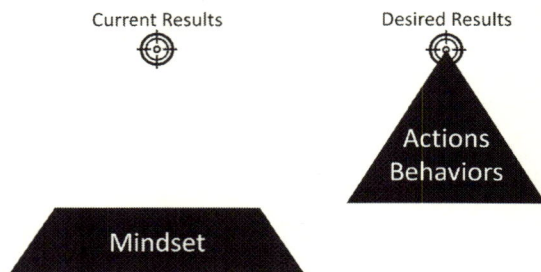

Figure 3.2: Attempting to change behavior only, adapted from the Arbinger Institute[34]

We start changing our actions and behaviors, but leave our mindset where it always was. And what happens? The mindset acts like a tether that pulls our actions and behaviors right back to where they were before, instead of giving us the sustained improvement we were looking for.

We can compare our mindsets to corrective lenses like glasses or contacts we use to see the visual world. Our mindsets are the mental lenses through which we see ourselves and the world around us. Mindsets determine how we interpret interactions, experiences, and relationships. Our mindsets, therefore, have massive impact that can either affect our lives positively or negatively. We can change some of these lenses or even update the "prescriptions" to help us see more clearly and improve our results over time.

If your "lenses" are clear, you are able to see life in its full, colorful reality. If not, you're going to have a distorted view of everything that happens in your life, and miss out on opportunities.

Your mindset influences your behavior, which then leads to the results you have in your life.

Susan was a client of mine who realized the importance of her mindset. She was laid off from her job of over 10 years and found out her family pet needed to be put down all in one day. Not a great day at all. Yet she recognized she had a choice—she could wallow in despair and see herself as a victim of her circumstances, which might lead to a lot of inaction, or she could properly acknowledge her losses while also identifying what she wanted to accomplish as she moved forward.

She chose to focus on the positive aspects of her situation, which served her well. This mindset helped her feel gratitude for all of the love she had felt with this family pet. She felt grateful that she received severance pay from her job which she was actually planning on leaving anyway, and that for a time, she wouldn't have work responsibilities. This would allow her to spend more time on personal development work and the career transition process she had been working on.

Her mindset fueled positive behaviors which led to better outcomes.

That positive approach to change looks more like this:

Step 1: Shift your mindset to align with the results you are trying to achieve.

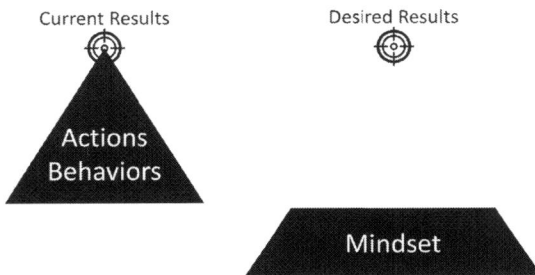

Figure 3.3: Changing mindset first, adapted from the Arbinger Institute[35]

Step 2: Your actions and behaviors will follow your mindset in a sustainable way, thus leading to the desired results.

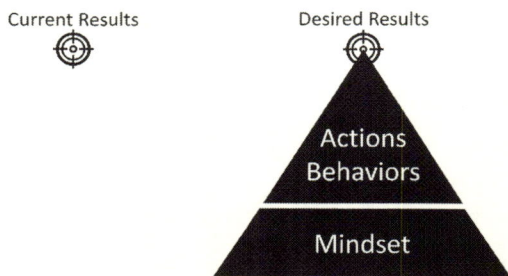

Figure 3.4: Changing mindset changes behaviors sustainably to produce desired results, adapted from the Arbinger Institute[36]

Shifting our mindsets first more sustainably shifts our actions and behaviors towards the results and goals we want to achieve.

GROWTH IS A CHOICE

In the fixed mindset, everything is about the outcome. If you fail—or if you're not the best—it's all been wasted. The growth mindset allows people to value what they're doing regardless of the outcome. They're tackling problems, charting new courses, working on important issues. Maybe they haven't found the cure for cancer, but the search was deeply meaningful. —Carol Dweck[37]

Probably the most well-known mindset we should work to develop in our lives is a growth mindset. It has been popularized and well taught by author and researcher Carol Dweck in her book *Mindset: The New Psychology of Success*[38] and her popular TEDx Talk.[39]

Let's start with the opposite of a growth mindset, which is a fixed mindset. In a nutshell, a fixed mindset comes from the belief that your qualities are carved in stone—who you are is who you are, period. Characteristics such as intelligence, personality, and creativity are fixed traits, rather than something that can be developed.

"Smart people succeed," says the fixed mindset. People with a fixed mindset believe that they are unable to significantly change, grow, or develop. Kind of sad, right?

I can relate to this. I have always struggled with art and other creative endeavors like writing, so I used to actively avoid opportunities to engage in these activities. They didn't come as naturally to me as math, science, and more objective subjects. I thought I was no good at it, so why bother trying to get better? Besides, in school, grading on those activities was completely subjective—why would I want to be beholden to someone's opinion like that? However, having this mindset meant that I didn't put any time into trying to improve, and as a result, I didn't get better. I regret that now, but at the time, I felt like I was held down by a boulder or rock, limiting my progress and opportunities.

A growth mindset, on the other hand, is the belief that your basic qualities can be cultivated through effort. Yes, people differ greatly—in aptitude, talents, interests, or temperaments—but everyone can change and grow through application and experience. Having a growth mindset leads to more effort and a focus on opportunities for learning, as well as a willingness to learn from mistakes. Even when we don't do well at something, we can try to learn from that experience.

"People can get smarter," says the growth mindset, "and do so by stretching themselves and taking on challenges."

Having a growth mindset doesn't mean that we don't lean into our strengths, but it means that we recognize that there's always room to learn and grow. It also allows us to give better feedback to other people because we believe that they aren't doomed to stay where they're at but can change too.

A growth vs. fixed mindset is demonstrated in many different situations, but one time it frequently manifests is when a person is trying to improve their career situation. Those with a fixed mindset might feel like they try and try and it's not working out for one reason or another. They might give up because they think if they're facing struggles as they try to progress in their career, they aren't good enough, smart enough, or something-else enough, and never will be. Why keep trying?

Alternatively, someone with a growth mindset will keep striving to learn, grow, and move forward no matter the outcome. They see even great challenges, setbacks, and rejections as opportunities. Consequently, they have more resilience and are more likely to achieve their goals!

If you want to develop a growth mindset, I invite you to identify a challenge you are facing and ask yourself, "What can I learn from this?" Even if you don't reach your intended goal, you don't need to view it as a failed experience, because you can still learn something valuable.

Another great strategy is to use the power of the word "yet." For example, if you struggle with art like I do, instead of saying "I'm not good at art," say "I'm not good at art...yet." This small change can open up your mind to the possibility of improvement.

This strategy was useful for one of my clients, Ahmad. He was an experienced mechanical engineer, but decided he wanted to lean into robotics. As such, he completed a rigorous Master's in Robotics while still working full time at a top startup.

But when we met, he was more than six months past his graduation and didn't feel like he had made any progress towards moving his career into a robotics role. There wasn't a great option at his current organization, and he was having a hard time connecting with others in the industry.

"I'm not good at networking," he told me.

"Yet," I said. "You're not good at networking...yet. But you can be."

"Okay, let's do it."

He put some significant work and practice into it, and within a couple of months, he had gone from having zero interviews to having eight interviews in one week, mostly because he was networking with people in companies that were doing things he was excited about.

Not bad for someone who wasn't good at networking...yet!

FOCUS OUTSIDE YOURSELF—AN OUTWARD MINDSET

Seeing people as people rather than as objects enables better thinking because such thinking is done in response to the truth: others really are people and not objects.
—The Arbinger Institute[40]

Many engineers and professionals I talk to care a lot about the impact they make through their work. Will their work make lives better? Does the company they work for promote social values they believe in? What is the overall societal impact of the work?

This is great! It's fabulous to desire impact beyond the paycheck and to work on something that provides meaning in our lives.

Yet sometimes we may be looking beyond the mark in regards to the impact we make. Does it really have to be just about the products and services we work on? What about the small impacts we make on those we work with, live with, and interact with on a daily basis?

And what if we could actually change the world just by changing our own mindset?

I believe we can, if we embrace an Outward Mindset, as taught by the Arbinger Institute.[41]

The Outward Mindset invites us to truly see people in our lives as people. When we do this, we treat them differently (usually much better). We care more about learning about others' challenges, hopes, fears, and dreams. We are no longer only focused on our own cares and desires. We spend more time focusing on the team's success, which encourages us to take actions that help others be successful.

The opposite of an Outward Mindset is having an Inward Mindset. When we have an Inward Mindset, we tend to be self-focused and view others as objects, which can include seeing them as obstacles in our way or just a means to accomplish our own objectives. We may use others to achieve our goals, or we may not help others even if we feel we should, because we don't believe they are worthy or we don't see them as people who matter.

One metaphor used by Arbinger to characterize the Inward Mindset is being "in the box" where we are only concerned with ourselves and cannot see others outside. It's all about us.

Think about what would happen if even just a few of us shifted to being more Outward most of the time. The good news is, an Outward Mindset in ourselves invites an Outward Mindset in other people. It spreads, and the ripple effect is vast.

It can be a great thing to think about products we work on that can impact thousands or even millions of people. Yet perhaps the greatest impact you make will be in the lives of the people closest to you that you are able to connect with in new ways simply because you shift how you see them.

Beyond the Golden Rule?

The Golden Rule, which is treating others the way we want to be treated, is a common concept that many of us have heard since childhood. But is that really the best approach to interacting with those around us? I suggest that there is an even better way to operate, inspired by the Outward Mindset.

A more advanced form of the Golden Rule, sometimes known as the Platinum Rule, says we should treat others the way *they* want to be treated. [42] We do this by considering *their* needs and challenges, and what *they* care about, not the way *we* want to be treated. Of course, this means we'll have to ask questions in order to understand what others want and how we can be most helpful. Considering others in this way is exactly what the Outward Mindset is about.

To shift from an Inward to an Outward Mindset, start by assessing your views of others and reflecting on your priorities. Some questions you can ask include:

- Do I view others as objects getting in my way? Or as a vehicle to help me get what I want? Or do I see them as people who matter like I matter?
- Is reaching my goals always more important to me than helping others reach their goals?
- Do I believe others are doing the best they can, and can I give them the benefit of the doubt when they act differently than I think they should?
- Do I genuinely care about others and frequently show it through my actions?

Additionally, you can identify people who make you feel your best and analyze what they do to show that they care about you. You can then make changes in your behavior to be more like them and cultivate an Outward Mindset. This shift in mindset can be life-changing and beneficial for both you and those around you.

"I Feel Like I'm a Better Human"

Marius, one of my clients, said this after we worked together for a few weeks to help him learn and apply the Outward Mindset.

Marius was an experienced civil engineer, but he wasn't progressing in his career as quickly as others seemed to be. In his industry, most engineers become project managers after 5 to 10 years of experience, but after 15 years, Marius was frustrated as he continued to wait for his promotion.

Marius and I met after he got a new job. He wanted to start strong and work his way towards becoming a project manager, but recognized that some of the ways he interacted with people at his previous firm didn't work well. He wanted to change, but he didn't know how.

Some of the limiting beliefs and mindsets we identified included:

- He thought if he took on new work that was unfamiliar to him, he would fail.
- He was afraid to speak up because he didn't want others to think he was stupid.
- He often put headphones on while working and completed tasks without interacting with others. Diving into being fully focused isn't a bad thing, but the problem was that he was nearly always shutting others out.

After learning about the Outward Mindset, Marius made several changes:

- He reached out to others to express appreciation for how they helped him and asked if there was anything he could do to help them.
- As he started his new job, he reached out to employees at the new company to get to know them. They responded positively and introduced him to more people which helped him feel more comfortable as he began the new role.
- He practiced being more aware of how his work impacted others.
- He spoke up in meetings because he recognized it was good for him and the whole team when he fully understood and contributed to the project and tasks.

These may seem like simple changes, but they were game-changing for Marius and dramatically improved the quality of his relationships at work. He was becoming a better human indeed, and he was on a path to become a great project manager.

THE PROCESS OF GROWING YOUR MINDSET

The first step in growing your mindset is knowing your current state and what the next step of change is.

Let's use a diagram to illustrate different states we can be in. Figure 3.5 shows a developmental sequence, often called the "Four Stages of Competence" that we go through as we work on changing our mindset.[43]

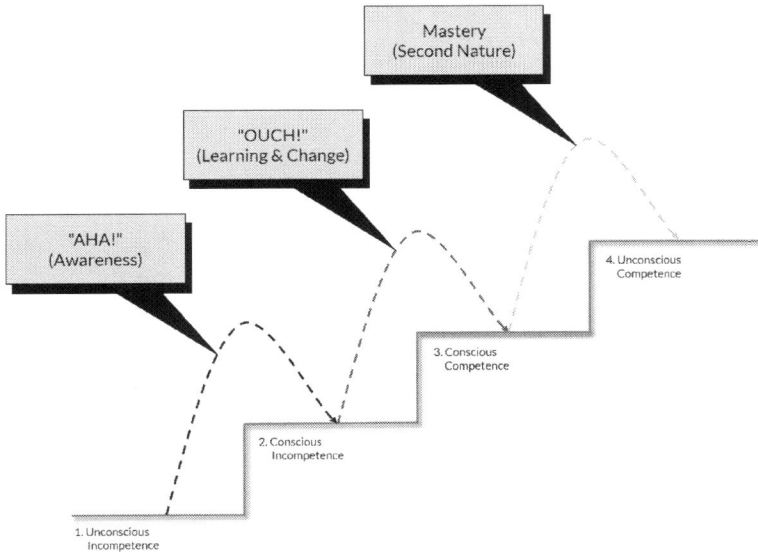

Figure 3.5: Four Stages of Competence, adapted from De Phillips, Berliner, and Cribbin[44]

According to the Four Stages of Competence model, when we are learning something new, or for the purpose of this book, developing a new mindset, we go through four stages:

1. **Unconscious Incompetence:** This stage could easily be referred to as "ignorance is bliss." We don't know what we don't know with regards to how our mindsets fall short. We reap the consequences from our unhelpful mindsets, but we don't understand why things are how they are. Unconscious incompetence can seem like a happy place to be until reality sets in and we become aware of what is really going on.

    Think about a kid who is about to start learning how to drive. Playing Mario Kart is the full extent of their driving experience, but they probably think they'll be an awesome driver. Will that be true? Probably not. Reality sets in at stage number 2.

2. **Conscious Incompetence:** At this stage, we are fully aware of how we fall short in regards to our mindset. We want to have a better mindset, but we come to realize that changing our mindset isn't as easy as flipping a switch. It can be frustrating to be at this stage, and this is the stage where many people give up. Don't let that be you!

    To continue our new driver analogy, this is the first time a new driver actually gets on the road and almost runs into another car.

3. **Conscious Competence:** Moving from Conscious Incompetence to Conscious Competence requires a lot of hard work. It takes continual conscious effort. But as we improve our mindset, our brain wiring changes—neural networks develop to reflect our new way of being. We are able to see improvement and gain proficiency, but our new mindset doesn't feel natural yet.

This is like a driver who, after a lot of hard work, can now drive well—they always look ahead to every stoplight, check every blind spot, and see every sign on the road—but driving well takes all of their mental energy.

4. **Unconscious Competence:** When we arrive at this stage, we have achieved mastery—our new mindset or way of being has become almost second nature. We can perform and operate at this new level without a lot of conscious effort.

   Our driver has worked and practiced and is now competent on the road without exerting extreme concentration. They notice and react to things as they come without having to think about it too hard.

So what stage are you now at with regards to your desired mindset? If you are this far into the book, you are at least at the Consciously Incompetent stage because you've become aware that there are some changes you need to make. And perhaps you're even further along than that.

No matter where you are now, you can take action and continue to progress. In the next chapter, I'll discuss steps you can take to shift your mindset.

TAKE INTENTIONAL ACTION

Take a Mindset Assessment

Where are you at with regards to your mindset? A simple way to gauge this is to take an assessment. Mindset assessment tools are imperfect, but provide insights to help you understand where you are and what progress you might want to make.

Here are a few assessments I recommend:

- The Personal Mindset Assessment, created by my friend, Ryan Gottfredson. This assessment is the most comprehensive mindset assessment I'm aware of. It rates you on four different mindsets (including the fixed/growth and inward/outward), and provides a lot of supporting material for improving your mindset. See https://ryangottfredson.com/resources/#assessments.
  I also highly recommend his related book, *Success Mindsets*.[45]

- The Mindset Works Mindset Assessment. This is a fixed/growth mindset assessment based on the work of Dr. Carol Dweck. See https://www.mindsetworks.com/assess.

- The Arbinger Institute Mindset Assessment. This is an inward/outward mindset assessment.
  See https://arbingerinstitute.com/MindsetWeb.html.

Examine Your "I Am" Beliefs

It's not always the people who start out the smartest who end up the smartest. —Carol S. Dweck

As mentioned earlier in the chapter, one of the most damaging and pervasive mindsets we can adopt is a fixed mindset. A fixed mindset makes you feel like you will never be able to change. Consequently, you might buy into many "I am" statements about yourself that are limiting beliefs.

I, too, have been guilty of believing negative "I am" statements. As a young aspiring engineer, I started believing in stereotypes about myself that limited my motivation to progress. A few statements I got stuck on included:

- I am book smart, not street smart.
- I am analytical, not creative.
- I am good with numbers, not with people.

- I am not good at things that are "subjective"—I want right answers!

These beliefs made me think I couldn't change or improve, holding me back from progress and achievements that could have been accessible if I believed in myself differently.

I invite you to shift your "I am" statements towards positive beliefs about who you are and who you can become. Repeating positive statements about yourself, day after day, will change what you think about yourself.

For instance, try saying, and believing, "I am an Intentional Engineer!"

To help me maintain a positive mindset about myself, every morning I write a few positive "I am" statements in my journal.

I invite you, also, to think of "I am" statements that resonate and feel helpful for you. Write them down. Post them on your mirror or fridge or put them in your cell phone where you will see them frequently. And when you start feeling down on yourself, repeat the statements to yourself.

Here are some "I am" statement I've been writing recently:

- I am a son of a loving God.
- I am a loving husband and father.
- I am an author (I took on that identity as I began writing this book).
- I am full of value to bring to the world.
- I am a good man.
- I am enough.

What "I am" statements will inspire and encourage *you*?

CHAPTER 4

SHIFT YOUR MINDSET TO CHANGE YOUR LIFE

We all have barriers or roadblocks to change. And we know if we could just get over them and turn our weaknesses into strengths, we would unlock all sorts of progress and opportunities. For Ahmad, discussed in the previous chapter, there were barriers with networking he needed to overcome. And for Marius, also discussed in the previous chapter, his barrier was shortcomings in his work relationships.

Before you can start making changes, you first need to know what it is you want to change.

For me, it's being more present with my family, my clients, and with my work. All too frequently, when I am with my family, my mind is somewhere else and I'm not fully present. When my wife notices this, she's great at bringing me back by saying something like, "Hey, be with us...." At work, I often lose myself in checking email or sports news updates instead of remaining focused on the work I'm doing. In fact, I was feeling pulled to do that right before I wrote this line. Yep, there is work to do!

I recognize that if I can continue to improve my ability to be present in all areas of my life, that will fuel significant progress for me.

Here are a few other examples from people I have worked with:

- John was an experienced engineer about three months into his new job at a space exploration company. He recognized that if he could effectively create better professional relationships, it would enable him to reach his goal of making a shift to work in a new group that is working on a project he is *very* interested in (think long-term living in space). But he was stuck in a routine of just taking on the tasks given to him with little to no focus on expanding his network or seeking new opportunities at his company.

- Prasha was a senior data engineering leader who had taken time off to get her health in check. But as she came back to work, she needed to figure out how to grow her team in a fast-moving startup while also staying healthy and creating boundaries for herself so she didn't burn out again. Basically, not taking it all on herself and saying "yes" to everything.

- Aditi was a senior program manager shifting to her first opportunity as a team leader. As a top-notch individual contributor, she was well-acquainted with delivering projects at a high level herself, but was new to the process of being responsible for a team. She knew she needed to take an interest in the work of those on her team and help them work together well in order to unlock the full potential of the group, but she wasn't quite sure how to make this shift.

GET TO THE ROOT OF THE ISSUE

So what's your big change? **What is the *number one* thing that would help you unlock your potential?** Take a moment and ponder, and possibly journal about this.

Once you have identified the change you need to make, it doesn't mean all the work is done—you're just getting started.

Even if you have known about the need to change (often for years), you might still resist it. You may have already tried all sorts of things including:

- Taking courses to learn new skills
- Setting up new behaviors/expectations (like a diet or exercise program)
- Reading self-help or leadership books
- Following a focused morning routine before starting your work day
- Trying to "buckle down" and make it happen this time

And yet, you still struggle to sustainably change the behavior. Why?

You are trying to prescribe a solution of new actions to take, without knowing the true cause of your problem. You think just laying out a set of behavior changes will make it all go away.

And that's just the thing—as discussed at the beginning of Chapter 3, it's not simply about behavior change. Because mindsets fuel behavior, remember?

- I couldn't just say, "I won't look at email updates during family time anymore" and think I was going to be more present.
- John couldn't just declare that he would network better and expect things to change.
- Prasha couldn't just expect herself to be able to easily say "no" to projects coming her way.
- Aditi couldn't just assert herself as a leader and wait for her skills to catch up.

For each of us, there is something beyond just our actions that is driving our inability to make these changes. **It takes a change in belief, a change in assumptions, a change in mindset.**

If you change your mindset, then suddenly you can address the real problem and get to a real solution.

If you don't, your efforts to change your behavior will feel like driving a car with your foot on the gas and the brake at the same time. It doesn't work.

To use another metaphor from Harvard psychologists Robert Kegan and Lisa Laskow Lahey in their book *Immunity to Change*, it's like your mind has an immune response to the change you're trying to make, and fights hard to keep things the way they are. In the book they say, "At the simplest level, any particular expression of the immunity to change provides us a picture of *how we are systematically working against the very goal we genuinely want to achieve*."[46] (emphasis added)

Wow, we are systematically working against ourselves? Yes, we are.

Let's take Ian as an example. Ian was a senior engineering leader, trying to build teams, build the business, and build his career while growing his young family (something he wanted to make more time for).

But he took it all on himself, without delegating and sharing the workload with his team. He was stuck in a mindset that exaggerated his self-importance, thinking he was the only person capable of completing the work correctly. Any new task that came his way, he thought he needed to take care of it on his own. So he quickly became overwhelmed by the weight of all his responsibilities, and wasn't able to spend the time with his family that he wanted to.

Don't get me wrong, he typically did a great job with the tasks he took on. He worked extremely hard and did great work. But crazy busy weeks piled up over time and he burned out.

He tried some things to figure out how to scale his impact without burning out again:

- He took courses on LinkedIn Learning.
- He attended industry conferences.
- He participated in training available to him through his workplace.

But none of these things seemed to do the trick. He was still stuck in his old ways of doing things. Months of effort and exhaustion with no end in sight.

The problem was, until we met and started working together, Ian didn't even know what the real problem was. Think about that for a moment!

But how do I uncover the real problem? you might ask.... There is a process for this, and I'll lay out the basics for you in a moment.

After working with me to make some internal and external changes, Ian later had a great experience where he told me, "I just took on a new project and was able to assign my team member to take ownership of it from the beginning. The client was cool with it and I hardly have to [do] anything about it. I *never* would have been able to do this a few months ago."

This was because he was able to get to the root of the issue. Ian eventually realized that he was driven by a fear that if he didn't take primary ownership of virtually all tasks that clients were counting on, the quality of the work would suffer and would result in poor performance. Ironically, trying to do it all himself actually made things worse. As he changed his mindset to being more trusting of his coworkers and oriented towards building the team, he started experimenting with delegating more and letting go of the emotional need to do everything himself. And as he changed, he received great results.

The real key to sustainable change is going through a mindset transformation. Shift your mindset, then let it shift your behaviors, which will bring the results you want.

STEPS FOR SHIFTING YOUR MINDSET

Here are four steps to make that shift:

Step #1: Write down the "big change" you want to make.

This is the thing that, if you shifted it, would likely make the biggest impact in your life and work.

Don't overanalyze it and wonder if you have written down the right thing. Just write it down. Paper is good. A sticky note is fine. Type it out if you want to do it digitally. Just get it out of your head.

For me, this is being more present. For Ian, discussed earlier, it was accepting that others on his team could do great work, that he wasn't the only competent one. For you it will be something different—that's what you need to identify.

Step #2: Write down all the fears you have about going through the process of making this change.

Here are some common fears about making this change that I've heard:

- The quality of my work will go down.
- I will get rejected by my boss.
- My family won't support me.
- I will try something very important to me and fail.
- My coworkers won't respect me.
- I will miss out on career opportunities.
- I might get fired.
- I might end up being financially ruined.

Whoa, that's a lot. These are just examples. What specifically are *you* afraid of?

Step #3: Write down an action you can take to test whether any of your fears will come true. Then run the test.

Yep, that's right, we're going to run an experiment and collect data. (We're engineers, right?)

Choose a small action you can take, but one that will require you to step outside of your comfort zone.

Here are a few examples:

- If you struggle with delegation, take one small task and train someone else to do it. Document the process and set them up for success (don't sabotage the whole thing from the start).

- If you are afraid to ask for new responsibilities that will expand your skill sets, look for a way you can add value in a new area that grows your capabilities and is also helpful to your team or company, and do that work. Then, the next time you are talking to your boss, let him/her know what you did.

- If you aren't networking and building relationships with people at work, set up a brief conversation with someone you already know but want to get to know better. Start there, and see how it goes. Prepare a few basic questions about things you want to learn about them and their experiences. Then ask them.

Step #4: Analyze the data.

After you've run the experiment, taking actions you typically would avoid taking, analyze what happened. Did all the terrible things that you thought would happen actually happen? If yes, maybe your fears have some validity (or maybe you sabotaged the experience and it was a self-fulfilling prophecy). If what you feared didn't actually happen, maybe there isn't as much reason to be controlled by those fears as you thought.

And suddenly, your beliefs and mindsets about the world are starting to change....

Going through these steps is like running the scientific method on your mind—you have a driving hypothesis that is leading you towards unhelpful behaviors, but you run a test and collect data to refute that hypothesis. As you do, you start to more fully believe that your fears are unfounded, and then they have less power over you! Here are ways I and some of my coaching clients mentioned earlier are implementing these "Steps for Shifting Your Mindset:"

- I am testing ways of letting go of my fear of missing something if I don't check my email constantly, allowing me to be more present with what is most important.

- John tried reaching out to a few new connections and learned that more people are willing to talk to him than he previously believed.

- Prasha said "no" to some new projects and realized that she didn't lose any influence in the workplace. In fact, people became much more discerning about areas that she and her team should focus on, and she got a promotion without burning out again.

- Aditi engaged in performance and corrective conversations as a new leader. Instead of receiving the pushback from her team that she feared, her team appreciated her caring approach to leadership.

As you go through this process of facing your fears and running experiments that disconfirm your beliefs again and again, your mindset will shift because you'll solidify more helpful neural connections in your brain. As brain scientists are known to say, "Neurons wire together, if they fire together."[47]

Even with a clear strategy for shifting your mindset as I have outlined, it's not like making a mindset shift is an instantaneous experience. It's a lifelong pursuit, but one that you can begin at any moment. Why not right now?

PROGRESS, NOT PERFECTION

This process of mindset change will take mindful effort and won't be free of bumps. It's normal to fall back into old patterns associated with your former mindsets. In fact, knowing that you've fallen back, and knowing how you can get going again, is a sign of progress!

Along the path of improving your mindsets, recognize that you are looking for progress, not perfection. We all slip into negative mindsets sometimes—it is impossible to be perfect. This news may be hard for you to handle if you're a recovering perfectionist like me, but it's true.

Brené Brown beautifully describes the limitations of perfectionism in her book, *The Gifts of Imperfection*:

> *Perfectionism is not the same thing as striving to be your best. Perfectionism is not about healthy achievement and growth. Perfectionism is the belief that if we live perfect, look perfect, and act perfect, we can minimize or avoid the pain of blame, judgment, and shame. It's a shield. Perfectionism is a twenty-ton shield that we lug around thinking it will protect us when, in fact, it's the thing that's really preventing us from taking flight.[48]*

Rather than perfection, we want to aim for healthy striving. It's good to work hard to improve and to have high expectations of yourself. But additionally, you must give yourself grace and self-compassion as you work each day to move towards improved mindsets, living your core values, and building relationships. Be patient with yourself if you don't progress as fast as you would like to.

Do your best, and recognize that your best is different each day. If you mess up or fall back a bit? That's okay. Congratulations, you're human! We all do it.

Becoming the person you want to be is a process, but one that I hope you enjoy even as you work through challenges along the way. You're giving yourself a gift as you progress on this journey.

TAKE INTENTIONAL ACTION

Steps for Shifting Your Mindset

If you have not yet completed the "Steps for Shifting Your Mindset" activity earlier in the chapter, do so now.

Change Your Questions, Change Your Life

One of my mentors, Richie Norton, taught me the principle: "Ask a better question, get a better answer."

The questions we ask ourselves about our circumstances and challenges have great power because the way questions are constructed influences the answers we seek out, believe, and accept as truth, thus influencing our mindsets. For example, if things in your life aren't going well in some way, it's easy to spend time asking "why" these things are happening to you, which will cause frustration and negative thoughts. If, instead, we ask something like, "What can I learn from this?" we will approach the situation more positively.

In her book, *Change Your Questions, Change Your Life*, Marilee Adams describes a tool she calls the "Choice Map." In the Choice Map, the questions we ask ourselves take us down different paths, and we can either choose our path or simply react to life.[49]

Some of the negative questions we might ask ourselves that put us into a state of reaction include:

- Whose fault is it?
- What's wrong with me?
- What's wrong with them?
- Why am I such a failure?
- Why does stuff like this always happen to me?
- Why are they so dumb?
- Why bother?

As you can imagine, questions like this never lead to helpful answers. I've learned that I can change from a destructive to a constructive internal dialogue by asking better questions! Here are a few questions I like to ask myself when I'm going through a difficult situation or a difficult time in my life. I hope these questions will be useful for you, too, to better empower you toward positive action when times are tough:

- What happened?
- What are the facts?
- What can I learn from this situation?
- What might they be thinking, feeling, and wanting?
- What am I responsible for?
- What choices do I have?
- What's the best path forward for me now?
- What is possible?

Changing the questions we ask ourselves is part of how we change our mindsets moment by moment, situation by situation. Asking questions like these when things don't work out how you would like will lead you toward positive action and help you be your best self.

So the next time you confront a challenge or frustrating experience, take a step back and consider the questions you are asking yourself. Write the positive, empowering questions above or other questions you like, and put them somewhere where you can easily refer to them. Ask yourself these questions as you create possible paths forward through your challenge.

CHAPTER 5

GET CAREER CLARITY

"I don't know what I want to be when I grow up." This is what my client, Kim, told me early in our work together.

Kim was an accomplished PhD engineer who had worked across multiple disciplines, but at the time, found herself in a boutique technical consulting role working for a narcissistic boss. She also happened to be a professional mountain biker on the weekend. But because of some of the frustrating challenges at work, her life didn't feel fun anymore.

She knew she needed a career shift, but what should it be? That was the big question we were working together to figure out.

She needed to find career clarity.

WHAT IS CAREER CLARITY ANYWAY?

Put simply, having career clarity means knowing what is most important to you in your career—knowing where you eventually want to end up. To establish career clarity, you go through the process of deciding *what* you want and *why* you want it, so that you can more easily create a plan for *how* to make it happen.

Having career clarity can help you decide if a company or role will be a good fit for you, what industry you want to work in, whether you should get more education or training, and more. Getting to a place of career clarity is essential before making important decisions concerning your career. Career clarity will be your North Star as you press forward. Once you have clarity, even if your path is winding, you can confidently navigate because you know where you want to end up. For example, if your goal is to work in robotics, you may consider taking a mechanical design job for a time but still study and improve your skills in controls, software, and automation to move towards robotic applications.

Being clear on what is important to you can act as a filter that eliminates different options that might seem appealing in some ways, but will work against your goals in the long run. Thus, having clarity narrows your focus and removes the pressure that comes from feeling like you need to consider all potential paths you *could* take.

When making decisions in your life and career, you can be confident as you lean on the decisions you have already made and act accordingly.

WHAT CAREER CLARITY IS *NOT*

One misconception about career clarity is that you have to figure out a step-by-step plan for the next 10 to 20 years of your career with exact detail. Let's be real.... Who can do that, especially when we work in a field that changes so rapidly, and things you will be working on in 10 years likely don't even exist right now?

No, you don't have to determine every single step along the way today. Getting career clarity is not like creating some magical treasure map that will show you exactly how to get from point A to point B, where X marks the spot on the map. Instead, your path will evolve as you go along. If you don't have your whole path figured out, don't stress—that is to be expected!

CAREER CLARITY WILL ACCELERATE YOUR ENGINEERING CAREER

Your vision will become clear only when you can look into your own heart.... Who looks outside, dreams; who looks inside, awakes.
—Carl Jung[50]

Getting clear on what you want in your life and the kind of person you would like to become accelerates your ability to be successful in your career. You are more likely to stay focused on things that are important and take more intentional steps toward growth and development rather than just letting your career happen to you.

To illustrate how a lack of clarity can slow you down, let's remember the classic tale, *Alice in Wonderland.* In the book (and corresponding movie), Alice finds herself lost in some dark woods in Wonderland. She comes to a fork in the road, and isn't sure which way to go. At this point she encounters the Cheshire Cat, who is kind of a sly character and not terribly helpful. Here is their exchange:

Alice: Would you tell me, please, which way I ought to walk from here?

The Cheshire Cat: That depends a good deal on where you want to get to.

Alice: I don't much care where—

The Cheshire Cat: Then it doesn't much matter which way you walk.

Alice: ...So long as I get somewhere.

The Cheshire Cat: Oh, you're sure to do that, if only you walk long enough.[51]

Perhaps you've felt like Alice before—knowing you want to go *somewhere* other than where you are, but also not really caring where.

Alice, in this instant, was the epitome of *not* being intentional. And the cat was able to point that out to her, even though he didn't do a great job helping her figure out which way to go.

The lesson, as it applies to your life and career? Figure out where you want to go, and then you can chart the path to get there and work to make it happen.

Sometimes we look to other sources for guidance on what we should do next in our career and lives. Certainly, mentors, guides, and coaches can be partners in this process (if they are more helpful than the Cheshire Cat). But in the end, the responsibility for getting clarity and living the life you want is up to you and no one else.

No one understands you better than you understand yourself. Only you have lived through every single one of your life experiences. So the person who has the greatest ability to unlock clarity in your life is no one other than you.

As you do as Carl Jung suggests and look inside, you can uncover some amazing truths about what you believe and what you are working to *become*. Having clarity does not mean being able to foretell the future, but seeing our past and present circumstances clearly, and deciding what kind of life you want to create. Who you are today doesn't have to be who you are tomorrow.

Once you're clear on where you want to go, you can start on your journey to get there! But how do you get clear?

HOW TO FIND CAREER CLARITY

Most people think they lack motivation when they really lack clarity. —James Clear[52]

Finding career clarity is not always easy, but it is possible. It takes time, effort, and self-reflection. But if you're willing to put in the work, you can find a career that is both fulfilling and rewarding. I have, on the pages that follow, provided a few specific actions you can take to go from confused to clear about what you want for your career.

Let's jump into it!

Get Writing

Often the simplest tools and actions are the most effective, but sometimes that means they are also commonly neglected. A journal is one such tool. From the beginning of this book, I've encouraged you to utilize a journal—if you haven't started yet, now is the time.

Writing in a journal can help you see a simplified view of your life, figure out what's most important to you, break through "analysis paralysis," and get the clarity and confidence you need in order to take action.

Many clients come to me with many possible options for where they could take their career. They have many skills and interests that could be applicable across industries and functions, but they don't know where to go next. The problem is that rather than taking steps to move in the right direction, the uncertainty often paralyzes them, making them unable to progress.

So in almost every case, I ask them to write. We move through some focused journaling activities, but it's also helpful to free write in a journal.

Go ahead and open a blank page and start asking and answering questions.

- What comes to mind when you think of your future life and career?
- What ideas do you have about which direction you'd like to head?
- What careers sound fun?

- What inspires you?
- What are you truly afraid of?
- What elements are most important to you in a career (such as company, culture, role, location, compensation, travel, in-person vs. remote, etc.)?

Answering these and similar questions will help you uncover insights you have been searching for, but, in many cases, may have been hiding inside of you all along.

As you work on finding career clarity, it's also helpful to create a list of "non-negotiable" items. Write down criteria that, if not met, immediately remove a role from consideration.

Get out of your head and write. Or, put another way, write to get things out of your head. And see what you discover.

Get Help Instead of Trying to Figure It Out Alone

Kedon was only about two years into his career and already felt stuck in a role he didn't like. He wanted more out of his job, but couldn't see a path forward—he didn't know how his skills and experience could translate to doing anything other than what he was doing. He had limited his thinking to only the immediate pain of the situation at hand. It is easy to get sucked into thinking patterns such as Kedon's.

It was hard to watch Kedon wrestle with his limited thinking as he failed to look at the big picture. But I, along with his friends and family, helped him discover that he, like all of us, has more potential than he realized.

When we deal with problems in our personal lives, careers, or other areas of our lives, it's so easy to get caught up in the pain and uncertainty of the moment. When we face these problems alone, it can be especially painful and debilitating. So rather than struggling through career clarity questions and challenges alone, reach out for help. Find mentors or coaches who can help lift your spirits, broaden your perspective, and help you find new opportunities. Enlist friends and family to support you. They can help you figure out what you want. (And once you have clarity, let everyone in your life know about your goals, which will put you on their radar as they interact with people that might be good for them to introduce you to.) Getting support from others is crucial to wading through the complexities of life, finding clarity, and maximizing your potential.

Additionally, the people who know you best in both your personal and professional life may have insights about you beyond what you would think of yourself. To get more clarity on the strengths they appreciate about you, you can ask questions like:

- What do you see as my greatest strengths?
- If you were to use three words to describe me, what would they be?
- What do you most appreciate about being friends (or coworkers)?
- What do you think I do well, but probably don't give myself credit for?
- What do I do that makes you feel confident in me?
- What are some specific examples of when I have demonstrated my strengths?
- How do you think I could develop my strengths further?

You can start with these questions, but always dig deeper to better understand the information they share with you. Of course afterwards, make sure you thank them for sharing with you.

Perhaps it feels awkward to ask people about yourself. But this exercise isn't about fishing for compliments. It's about seeking support and insights from those who know us best. If you feel awkward asking someone these questions, just consider—if they asked you to share these kinds of insights about them, would you be happy to help? If the answer is "yes," they likely feel the same way.

Don't limit yourself by trying to shoulder the burden of reaching your goals all on your own—lean on others to help you reach higher. Another way you can ask for help is by asking for directions when you lose your way.

Ask for Directions

Have you ever been traveling to a new place and found yourself unsure of how to get to your destination? This is one of my pet peeves, especially when I'm the one driving.

When you're lost, somehow, some way, you need to get some help to figure out where to go.

In modern times, we often turn to our phone GPS (which isn't always 100% accurate in my experience). But if we are in a place with no service, we might *gasp* actually have to ask another human for help!

When we're unsure of where to go to find a desired destination, we need to ask for directions. Likewise, when we're lost or unsure on how to navigate our career, we need to ask for guidance. Asking for help from others as you navigate the uncharted territory of career decisions helps you obtain important insights from others who know the territory better than you.

How often do people respond negatively when you ask for directions? Virtually never, and if they do, it's probably because they are just having a bad day.

Asking for help doesn't need to be complicated and formal. The people you talk to will see your earnestness, curiosity, and passion. This will increase your confidence and at the same time improve your opportunities and insights.

Reduce your stress as you reach out to get insights from others by simply "asking for directions." This will open opportunities in your pursuit of success in the field you desire in engineering or beyond.

Make Decisions, Not Just Goals

It is in your moments of decision that your destiny is shaped.
—Tony Robbins[53]

There is a *huge* difference between creating a goal and making a decision. A goal is mostly just an idea. It's something you want, something you aim for.

A decision, on the other hand, is something you are fully committed to. You are willing to make sacrifices and do uncomfortable things to make it happen.

Making a decision requires you to move from just having a goal to taking action. You are no longer just hoping for something to happen, you are making it happen.

If your future is uncertain, it might be tempting to stay aloof from your goals so as not to be disappointed if things don't work out. I suggest doing the opposite. Decide and commit. If you don't, you're actually likely to inadvertently self-sabotage your situation to make your goals *not* happen because you don't really give it your all.

But if you fully commit to a path, you'll give it a full chance. Then, if it doesn't work out, you can know for sure that it wasn't the right path (at least at that time), and can pivot accordingly.

Are you willing to turn your big goals into decisions?

Take Action Despite Uncertainty

In the absence of clarity, take action! —Philip McKernan[54]

When dealing with uncertainty, we have to fight against the natural psychological and biological processes that work against us and make us feel unable to take action. Often when we have less information to go on, we make more irrational and erratic decisions. This is because when we don't have much information, we're more likely to feel fear, and that fear shifts control to the limbic system in the brain.

Making irrational and erratic decisions can sometimes work great as a survival mechanism (like when cavemen had to respond to unknown threats hiding in the bushes), but not quite as well in dealing with modern uncertainty about how our choices may or may not deliver the outcomes we are seeking.

There is no way to know how any path will turn out. But if you stay stagnant, *nothing* will happen and you will continue to lack career clarity!

So you have to be willing to take action despite your lack of clarity. In doing so, you'll actually try more things, increase your learning, meet more people, *and* get more clarity in the process. Essentially you're collecting more data, which is not possible unless you take risks and try new things. And yes, you might end up on the wrong path for a time, but sometimes ending up on the wrong path is the fastest way to get clarity about which path is the right path. As Jeffrey R. Holland said, "There are times when the only way to get from A to C is by way of B."[55]

No matter what, you can get something positive out of any experience. If you end up loving it, great! Double down and do more of it! If not, that's fine too. You can be grateful for learning from the experience and change directions for the future.

Don't let fear drive you to inaction. Move forward with intention.

Be Flexible and Patient

Stay committed to your decisions, but stay flexible in your approach. —Tony Robbins[56]

As we work towards career clarity, if we are flexible and patient, we tend to figure things out eventually. As I started my career, I had certain ideas about what I wanted to do, but as I was flexible, I found a different path that was right for me. For many years, I had the goal of earning an MBA degree. I wanted to balance my technical knowledge, education, and experience with greater exposure to leadership, business strategy, finance, and marketing.

About two years after my undergraduate degree and still in my first job out of school, I felt the need for a change—I was bored and felt stagnant—I didn't feel like I was learning or progressing most of the time. I thought that perhaps it was time for me to pursue the MBA degree to propel me towards the next steps in my career.

I was ambitious, so I started applying for top schools like Harvard, Stanford, Northwestern, and MIT. It wasn't long before I started receiving rejection letters, but there was one school that was willing to give me an interview.

In the midst of the application process, a new job opportunity came up. It paid significantly more than my current job and would give me new opportunities such as learning business principles on the job, leading teams, and developing new products, all things I wanted to try.

Having multiple good paths to choose from created a conundrum for me. I was excited about trying my hand at a top-tier MBA program (something I had dreamt of for years), but this role was exactly the type of role that I was hoping earning an MBA would allow me to explore in the future.

In the end, rather than continuing to test the waters of MBA programs, I took the job. But I never lost sight of my dream to get an MBA. A few years later, I decided to pursue an MBA from the University of Washington—they had a great program that was flexible and mostly online.

This allowed me to continue working and even start my business without having to move to a new place or incur crazy amounts of debt. And the things I learned and the people I met were invaluable. I was flexible with my approach, but I still fulfilled my dream.

Often it takes time to realize our desired future. Big things rarely happen in an instant, but we should never stop working towards our vision.

BACK TO KIM

Kim, the PhD engineer who I mentioned at the beginning of the chapter, used every single one of the techniques we've just discussed. In the end, her path to making a career shift was a shift that, previously, she didn't even think was possible. Here is just some of what happened:

- Kim found deep motivation to be more intentional because she realized that her parents had both hated their professional lives when she was growing up. She didn't want that to be her reality.

- As she spent time writing and reflecting in her journal, she gained clarity on what she really cared about in her work and in areas outside of her work. Then she focused on finding a career opportunity that would support those things. She realized that she loved interdisciplinary technical work, but also loved leadership, working with teams, and creating solutions collaboratively. It was hard because she didn't fit into a typical "scientist" or "engineer" position, which she was okay with, but for some people was hard to explain while looking for opportunities. She was committed to finding a role that fit her desires, rather than trying to fit someone else's mold just to get a new job. She made a decision.

- Kim became interested in technical project management, which was a perfect match for her blend of skill sets, and something she didn't know existed previously. To help her develop skills in this area, she started working towards earning a Project Management Professional (PMP) certification. This took some time, but she knew it would pay off. She was flexible and patient.

- She learned that her mountain biking friend worked at a prestigious national laboratory. They passed her resume along to project management teams who were enthusiastic about the possibilities of working with Kim. She got help instead of doing it alone.

- Through a chain of contacts, Kim ended up interviewing at the national lab for a role that was custom-made for her. She took action, despite uncertainty.

- The new job resulted in an immediate 40% raise, and a move to a region that was less urban, excellent for mountain biking, and had less expensive housing—all things she wanted.

- I caught up with her a year and a half later and she told me she had already received three raises, two of which were promotions.

Pretty cool what happens when you get clarity on what you want and work to make it happen.

TAKE INTENTIONAL ACTION

Go Deeper with the 5 Whys

If you have your why for life, you can get by with almost any how.
—Friedrich Nietzsche[57]

Many engineers have used the 5 Whys[58] tool to get to the root cause of a problem in their work. It's a tool that was originally developed by Sakichi Toyoda, the Japanese industrialist who founded Toyota Industries, and it became particularly popular in the 1970s.

It's a remarkably simple concept. We start with a problem that has occurred and drill down to its root cause by asking "Why?" five times. This helps us get beyond surface-level issues and solutions to figure out what's really behind the problem.

In addition to using it as a root cause analysis tool, I have found that using the 5 Whys can help you dig deeper into your "Why" (your deep, intrinsic motivation for reaching your goal), and help increase your career clarity.

Ready to use the 5 Whys? Get out your journal, and let's do it.

First, think about a big career goal that you have. Then ask yourself (and journal about) this simple question:

Why is [insert goal] important to me?

Answer with the first thing you think of. Do not get too complicated. Take that answer and ask the same question using the new answer.

For example, if you said "make $100,000 per year," you would then ask:

Why is "making $100,000 per year" important to me?

Your answer might be something like, "Because it's important for me to comfortably provide for my family." You then put that answer into the next "why" question.

Why is "providing for my family" important to me?

Because I want my children to have great opportunities in life.

Why is "giving my children great opportunities in life" important to me?

Because I want my children to have a better life than I had.

Why is "giving my children a better life than I had" important to me?

Here, you may uncover something about your past which is driving your decisions in the present.

Continue on this path for at least five levels. 5 Whys. And if it's beneficial, go even deeper, to seven or eight whys. Go as deep as you can. This activity will help you explore key life events that have shaped you, and important beliefs and values that can help you get to the core of *why* you are doing what you're doing.

You can also explore multiple paths of this exercise using a mind map. You may have multiple branches as different answers to each question, and can go deeper on each one of those. Have some fun with it, and see if you can find a pattern to discover your "Why."

Explore All Areas of Your Life

When working towards clarity and making goals, you need to look beyond just your professional or career aspirations. All facets of life are connected. In Stephen R. Covey's *The 7 Habits of Highly Effective People*, he identifies four dimensions in our lives: physical, spiritual, mental, social/emotional.[59] Here are some suggested activities to help you explore these dimensions in your life. It's a good idea to record these activities in your journal.

- For each dimension, physical, spiritual, mental, and social/emotional, spend time brainstorming goals and dreams you would like to accomplish.
- Consider why each of these goals and dreams is important to you. (You can use the 5 Whys activity to do this.)
- Identify the most important goal or dream in each dimension. Most important *to you*. Think about what it will mean to you once you have accomplished it.
- For each of these top goals (one in each dimension), what actions are necessary in order for you to accomplish them?
- Spend time imagining or visualizing what it will feel like when you have accomplished these goals. Who will you have become?

Assess Your Willingness and Commit

In order to be able to make the changes in your life you want to make, you have to be willing to do what is necessary. This means making a commitment, or being truly dedicated to the cause or goal you are striving for. One of my favorite quotes is from Richard G. Scott: "To reach a goal you have never before attained, you must do things you have never before done."[60]

Are you willing to do things you have never done before in order to reach your goals and dreams? This will take courage and determination to push through fear, uncertainty, and challenges. It will require sacrifices of time and energy to make it happen. It will take a full commitment to the process.

As you consider your willingness and desire to commit to making your goals happen, spend time answering these questions in your journal:

- What sacrifices will I have to make to achieve my goals? Am I willing to make those sacrifices?
- What will I have to give up? Am I willing to do this?
- What am I afraid of?
- What would be the cost of not taking these actions, and therefore not accomplishing my goals?
- How will achieving my goals change my life?
- What actions am I willing to commit to in order to accomplish my goals?

Then, recall or refer back to the core values you identified in Chapter 2. Do your answers to these questions align with your values? If not, reassess and rework your answers. If your answers do align with your values, resolve to follow through on your commitment, whatever sacrifices it may require!

CHAPTER 6

DISCOVER YOUR GENIUS ZONES

When I first met Shankar, he was a successful senior engineering leader at a media company. He loved building up his teams and delighting stakeholders, and had accomplished some huge project deliveries including cloud transformations, managing dozens of website migrations to new systems, and leading technology strategy.

The interesting thing was, he had a hard time talking about his strengths and what actually made him unique compared to anyone else. That is, until he learned about the concept of genius zones. I'll share more about Shankar's story later.

The idea of a genius zone is that each person has a special combination of background, expertise, attributes, talents, and passions that differentiates them from most, if not all, other humans on the planet. For instance, in addition to your general engineering skills, you might be an expert in a niche area of engineering, good at writing, and skilled at diffusing tense situations. Combining those skills reveals one of your genius zones.

There are many people who have each of those individual skills, but not many people have all of your skills combined (whatever your skills are). Your genius zone differentiates you from others and allows you to make a unique contribution at work, at home, and in the world.

Let's say you also speak Chinese, have experience creating products, and have a knack for bringing order to chaos. This combination of skills reveals a second genius zone. By combining all of your skills, attributes, and talents in different ways, you have many different genius zones.

When you choose to do work that allows you to fully apply one or more of your genius zones, you are able to do your absolute best work.

One of my favorite books on the subject of genius zones is *The Big Leap* by Gay Hendricks.[61] In the book, he actually describes four different zones we can find ourselves in:

1. **The zone of incompetence**: If you are doing work in this zone, you are engaging in something you are simply not good at. This is like me trying to create a clay sculpture. I have no experience with anything other than Play-Doh, and that's never pretty.

2. **The zone of competence**: You can get the job done, but not really any better than the next person.

3. **The zone of excellence**: In this zone, you are doing something you are very, very good at compared to most other people. This is what most people would call a strength for you.

4. **The zone of genius**: When you're in your genius zone, you are in what author Mihaly Csikszentmihalyi calls "flow."[62] (More on flow later in the chapter.) You're taking advantage of your natural talents and doing what you

earnestly enjoy. You have what could almost be called an "unfair advantage" over your competition.

Alternatively, the author and speaker Laura Garnett defines it as a simple formula:

Your Innate Talent + Greatest Passion = Zone of Genius[63]

However you choose to define it, discovering and identifying your genius zone will help you get clear on what work is best for you to engage in.

Even more important, this clarity will help you move *beyond* spending your time in the zone of competence and zone of excellence and into spending more time in your genius zones where you will do some of your best work *and* enjoy it!

WHY YOUR GENIUS ZONES MATTER

I think of the Zone of Genius as a continuous spiral. You go higher and higher every day as you expand your capacity for more love, abundance, and success. It's an upward journey with no upper limit. —Gay Hendricks[64]

All of us have genius zones whether we recognize them or not, and we are also constantly creating and expanding our genius zones as we learn and grow.

Ideally we'll spend as much time in our genius zones as possible, but sadly, many of us rarely if ever create the opportunity to make that happen. We find ourselves getting stuck doing things we are good at (tasks in the zone of competence or even excellence) instead of engaging in our best work by utilizing our genius zones.

This can be an easy trap to fall into, as you likely have many skills and areas where you are competent or even great. These areas might be your zones of excellence. But your genius zone is next level, and allows you to utilize your very best abilities!

I noticed that when I had opportunities to teach and train others, I was not only good at it, but I loved it so much that I hardly noticed time passing. After realizing that, I combined that skill and interest with my wide engineering experiences, and doing coaching and training with engineers became a true genius zone for me!

Doing work within your genius zone makes it easy to focus. You will have to let go of some work and tasks that you aren't well suited for (and maybe some that you are well suited for), in order to focus on the tasks you are *best* suited for.

Everything you have done and might do in the future can lead you to identify your genius zones and work within them as much as possible.

CHALLENGES TO FINDING YOUR GENIUS ZONES

Finding and living more in your genius zones isn't always an easy process. It's helpful to recognize the challenges to this process so you can diminish their power. There are three primary challenges to identifying and applying our genius zones:[65]

1. **You're not objective about yourself.** It's impossible to fully separate your emotions and assumptions about your experiences from the pure facts. You can't be a "fly on the wall" or see an outside perspective of your skills and strengths, which makes it hard to understand your specific and unique approach to work and life.

2. **You've been conditioned by your environment.** You only know what you know because of the experiences you've had and the influences you've been exposed to. Many of your views and opinions about what makes a great career path are likely influenced by your family, community, and

society. This is why so many kids only think about a few career options, like firefighter, doctor, or astronaut. You may not even be aware, yet, of roles that exist that would allow you to fully apply your genius zones.

3. **You want social validation.** Sometimes people choose a career path, whether consciously or subconsciously, in order to please others, without any thought to their genius zones. Of course everyone wants to be liked and accepted by their family, friends, and others they interact with, but should that be the reason for making decisions about our lives and careers?

These roadblocks, while challenging, are not so powerful that you can't overcome them. Let's jump into how to unleash the power of your genius zones.

UNLOCK YOUR GENIUS ZONES

When you identify and lean into your genius zones, you're really unlocking your potential. It enables you to do so many things that enhance your life and career, including:

- Do work you love, thus increasing satisfaction
- Improve your performance at work
- Communicate your greatest strengths to your current or potential employers, thus advocating for yourself and showcasing your value
- Impact the industry and other people in unique ways
- And so much more!

Now for some specific information to help you uncover your own genius zones and then utilize them more fully in your life!

Find Your Flow

It is when we act freely, for the sake of the action itself rather than for ulterior motives, that we learn to become more than what we were. —Mihaly Csikszentmihalyi[66]

When you do work that fully applies your genius zones, chances are, you completely lose track of time. What activities make you lose track of time? At work, it could be speaking to a group, or working on a challenging project, or solving a complex technical problem. Outside of work, it could be playing basketball, playing an instrument, or perfecting your latest recipe—anything where you have so much fun that time flies.

That feeling or state can best be described as "Flow," a term coined by the late researcher *Mihaly Csikszentmihalyi.* (I highly suggest you read his books and watch his TED Talk[67].)

Doing activities that put us in a state of flow increases happiness, fulfillment, productivity, and so much more. It's amazing. Identifying what puts you into a state of flow is a good first step in identifying your genius zone.

One of the key ideas from the research is that in order to find flow, we need to engage in activities that challenge us but that we have the skill or ability to accomplish. If a task is too challenging for you, or too easy for you, it won't be enjoyable.

To use the example of playing basketball—a high school basketball player would have no fun playing against a five-year-old (except that it's cute), but also would get torched by an NBA player and thus be disheartened. A brand new aerospace engineer probably wouldn't be in flow if they were asked to design an entirely new passenger airplane from top to bottom, but also would get bored if they were tasked with designing a paper airplane. This concept is illustrated in figure 6.1.

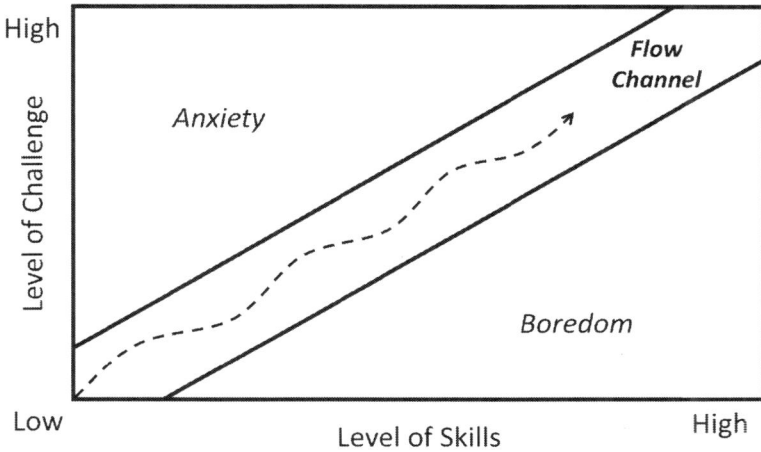

Figure 6.1: Flow channel diagram, adapted from Mihaly Csikszentmihalyi[68]

Knowing the importance of choosing tasks with the right level of challenge, if you are in a situation where you are rarely, if ever, being challenged, then you are likely rarely in flow, and therefore need to take on tasks that are more challenging. And if you often take on tasks that you are completely unprepared to do, you also likely are rarely in a state of flow. You can either take on tasks that you are better equipped to accomplish or you can increase your abilities.

Reflect on moments when you have felt flow in the past. What were you doing? Who were you doing it with? What was it about the environment or situation that made it work for you? Knowing these things, consider more—what possible actions can you take to easily access more moments of flow? And once you are in the flow, how can you keep it going? Recognizing flow experiences will help you do the work you are best suited for and will be most enjoyable to you.

Look For Unique Combinations

To find your genius zones, reflect on and identify combinations of your natural gifts, experiences, and learned skills that, when combined, create your unique approach to life.

These three areas have a distinct ability to feed into your genius zones. Let's explore further.

Natural Gifts

Yes, we all have certain strengths that we were born with. These are strengths that come naturally to us. They are different for each person, and that's a good thing, but it's worth identifying them for what they are, and you can call on these gifts in times when you need to use them to reach your goals.

You may have natural tendencies for math, science, and problem solving that led you to becoming an engineer in the first place. Excellent! Write them down. But don't stop there. Look for things that have always come easy to you in all facets of life, or at least types of work and activities you have enjoyed from the beginning, and write those down too.

Experiences

Our experiences shape our lives and create capabilities we otherwise wouldn't possess. Look back on childhood experiences, family environments, physical or mental disabilities you dealt/deal with, and more.

For example, I grew up in a religious faith (The Church of Jesus Christ of Latter-day Saints) that encourages and provides many public speaking opportunities. From the age of five or so, I prayed, read scriptures, gave short talks, and sang in front of dozens of people and sometimes our entire congregation of hundreds of people. Later, I had the opportunity to serve a full-time mission for two years where I constantly met new people, trained individuals and groups, and spoke to people about very personal life beliefs and experiences.

All of these experiences shaped me as they helped me be more confident and comfortable with public speaking, networking, and even doing coaching work where I get the chance to help people improve areas of life they care deeply about.

What experiences have you had that have shaped you and given you perspectives and expertise you otherwise wouldn't have? Write them down.

Learned Skills

Throughout your life, at home, in school, at work, serving in your community, etc., you've learned many skills. What are they, and how can they serve you now?

These skills can include technical skills like analysis, software languages you can write, etc. It can include personal or "soft" skills like your ability to run a great meeting, give a presentation to a diverse group, or communicate a strategy to a team.

These are skills that didn't just come to you, you have had to work for them. You learned from experts. You went to school or took courses. You read books. You received degrees and certifications. You have spent hours practicing and honing your craft. What are the top skills that have come out of all of that work and effort you put in? Write them down.

(Note: you're likely to find there is a lot of overlap between your natural gifts, experiences, and learned skills. That's to be expected—just choose the area where you think a skill fits best.)

Bringing It All Together

Once you have identified your gifts, experiences, and learned skills, you can look for ways to put them together in unique combinations to create genius zones! It's great to draw a Venn diagram (like the one in figure 6.2) for yourself and look at where your various gifts, experiences, and skills intersect. You can combine two, three, or ten different elements to find something that makes you unique!

For example, I have experiences working in and with engineers across disciplines including mechanical, software, electrical, civil, and more. I have some natural abilities for emotional intelligence that I've grown through practiced experience in having impactful conversations with many people (like in my missionary experience). I've also learned and practiced skills in one-on-one coaching, group facilitation, and training on various topics. See figure 6.2 for a Venn diagram illustrating this genius zone.

Putting some of my areas of expertise together, you can see that my genius zone is the area where they all overlap. This particular genius zone enhances my ability to train, coach, and facilitate leadership and career development work with engineers and technical professionals.

At the beginning of this chapter I introduced Shankar who knew he had the potential to make a big career upgrade. He wanted

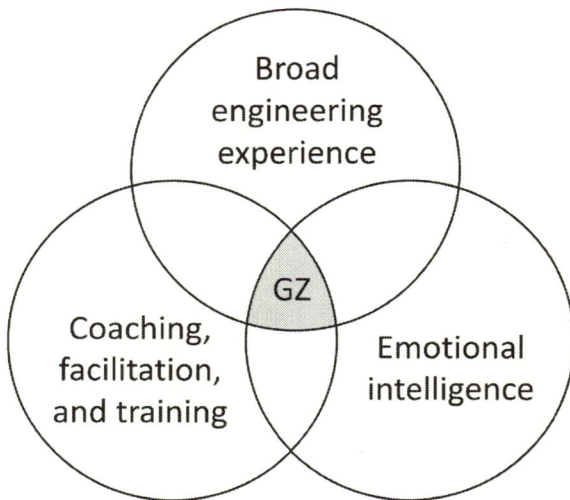

Figure 6.2: My genius zone Venn diagram

to explore career options that would allow him to fully apply his genius zones. And he needed to be able to communicate about his genius zones to potential employers. The first step was to identify his genius zones.

Working with me, Shankar identified technical strengths like cloud development, data engineering, and creating repeatable engineering systems. He also has excellent non-technical strengths like being skilled at empowering his teams, and an ability to bring together product, engineering, and business groups under a common vision and plan in order to execute on big initiatives. See figure 6.3 for an illustration of one of Shankar's genius zones.

After identifying this and other unique combinations of Shankar's skills, we discussed past experiences Shankar had had that illustrated how he had successfully utilized his genius zones. Then we took those experiences and crafted stories he could weave into networking conversations and job interviews to showcase his strengths.

Figure 6.3: Shankar's genius zone Venn diagram

Even though these genius zones were a part of Shankar, something inside him and part of his history, he had never explicitly identified them or learned how to communicate them to others. Doing so helped him feel more confident as he interviewed for leadership opportunities at some of the largest tech companies. Shankar's stories gave these companies a chance to see how he could do great work for them.

In the end, he landed a new role at a great company, almost doubling his salary. He continues to thrive as he builds global teams solving big problems. Identifying his genius zones didn't just help Shankar get the job, but continues to help him lean into making the contributions he can best make, while delegating other activities to those on his team.

Are you ready to find your genius zones?

TAKE INTENTIONAL ACTION

If you haven't started, it's now time to get serious about uncovering your own genius zones that you can utilize as you build a life of intention. Here are some activities to help you do just that!

Reflect on Flow Experiences

Earlier in the chapter I challenged you to think about times when you've been in flow because activities that put you into a state of flow are likely part of your genius zone. If it's difficult for you to come up with anything, start simpler. Make a list of the activities you enjoy and are good at in your work and other facets of life. Don't filter anything yet. Set a timer and write for at least 5 to 10 minutes.

After your time is up, look at each activity on your list and consider how you feel when you are engaged in these activities. When do you feel most "alive"? When are you your best self? Have you noticed that while doing any of these activities, you tend to lose track of time because you are fully in the moment and having so much fun?

This is a start to discovering when you experience flow. But keep going!

Go another level deeper and consider what it is exactly that gets you into that state of flow. Is it the activity itself? Is it the environment? Is it the challenge you're facing? Something else? Look for patterns and clues. The more details you can identify and replicate, the more you can use that information to focus on these activities in the future. Consider designing some specific experiences to get you into flow in the future. Then try them out and see how it goes.

Combine Your Strengths to Identify Your Genius Zones

Start exploring combinations of your natural gifts, experiences, learned skills, passions, etc. (Draw Venn diagrams if it's useful.)

Do any of these combine in ways that make you unique? Don't limit yourself to exploring work-related skills. Look at *all* areas of your life!

Consider if there's one skill or attribute that *must* be included (your engineering prowess, for example). You can consider this one your key area of expertise. You can then combine this expertise with any of the other items on your list. Perhaps it's just a combination of two areas, but if you have three to four skills and experiences that come together in a cool way, that's amazing, and probably something few people have!

Next, if you're considering utilizing your genius zones to make a career transition, search online and begin having conversations with professionals that are in occupations that might best utilize your genius zones. Doing so will help you validate or disprove your assumptions about these potential career paths and give you insights into how your genius zones could serve you well!

Ask Those Who Know You Best

As mentioned earlier, one of the challenges to uncovering your genius zones is that it can be difficult to be objective about yourself. In order to address this tendency, it's wise and helpful to ask others for insights about you and your unique strengths.

Pick five people who know you well, especially in a work setting, and ask them these two simple questions (adapted from Laura Garnett):

1. How would you describe my approach to work and how is it unique?

2. As we work together, what things do I do that create the greatest impact?[69]

As always, dig in and get curious beyond the initial answers and ask follow-up questions. Getting this insight can help you begin to uncover some of your innate talents and ways you work that set you apart, and can also help you identify areas where you would like to improve.

What New Genius Zones Do You Want To Develop?

Genius zones aren't static. They are always growing, changing, and developing because you are!

Now that you've identified some genius zones through looking at flow experiences and unique combinations of strengths, as well as asking input from coworkers, it's worth considering if there are genius zones you still want to develop to help you achieve your intentions and big goals. What new genius zones would help you become your best self? In order to have the impact you want to have, you can't stay the same, and part of changing to make a greater impact can include intentional work to create these new genius zones!

It's possible you'll be able to develop some of these new genius zones simply by combining expertise you already have in unique ways that you haven't already. In other cases, developing these new genius zones will require new expertise. What will you need to do to obtain that expertise? You might need to complete new degrees or certifications, put yourself out there to try out some new experiences, request on-the-job training, or hire a coach or mentor who can help you develop or enhance a skill.

As you reflect on what genius zones you'd like to develop, make plans to begin today. Record your intention in your journal.

CHAPTER 7

CONTROL YOUR NARRATIVE BY GROWING YOUR PERSONAL BRAND

Brenden reached out to me at a point in his life when he felt stuck and didn't see a career path that was compelling to him. He sent me this note:

> *I am currently working as an applications engineer in the additive manufacturing industry, but I am attempting to transfer to a more technical role where I can further my career...and fully utilize my degree and the skills that I have acquired. Namely, I am attempting to obtain either a more technical AE role, a mechanical/design engineer role, or a manufacturing engineer role.*
>
> *Your services caught my eye, as I have been trying to make this transition for some time, but have not had much luck.*

As we talked, there were two things that stood out for me that Brenden needed to work on:

1. Getting more clarity on what he really wanted to do next. (We utilized tools discussed in Chapter 5 to help him do this.)
2. Creating a personal brand that allowed him to tell his story and showcase his value.

As an applications engineer, Brenden was a problem solver who worked very closely with clients, but never got deep enough into projects to truly solve significant technical issues. Mostly, his work was focused on increasing sales of devices and software to clients.

Because of this, it was hard for Brenden to tell a compelling story and show his capabilities to do more technical work for a potential new employer.

He needed to create his own personal brand.

WHAT IS A PERSONAL BRAND?

Everyone has a personal brand by design or by default.
—Lida Citroën[70]

What is your personal brand? Put simply, your personal brand is the way you present yourself to others that influences how they perceive you. You want others to evaluate you on your expertise, experience, and personality. Your brand is created through a combination of what others already know about you; their impression of you based on what they learn about you online, based on what they hear about you from others, and through interacting with you in person; and more.

Others will, in the end, make up their mind about who they think you are based on multiple data points. Your task is to help guide them to understanding more of the truth about who you are, and how your expertise can benefit them (especially in a career context).

For some engineers, the thought of "personal branding" might seem cringeworthy, but personal branding in the context of career development isn't the same as creating a brand you can market products for. No need for Super Bowl commercials here.

As the Citroën quote at the beginning of this section suggests, you have a personal brand whether you like it or not—so wouldn't you rather intentionally create it?

Think about it this way: The people in your life who are close to you know you well because they spend a lot of time with you. However, your personal brand is important for those who only know a little bit about you. They only know what you tell them, so what do you want them to know?

Imagine a simple Venn diagram, like the one in figure 7.1. In one circle are all of the attributes, skills, personality traits, interests, and experiences that make you who you are. In the other circle are all of the ways you show who you are to other people—online materials, your resume, what you share on LinkedIn and your personal websites, others' experiences interacting with you, and more (labeled "what people know about you" in figure 7.1). The intersection of the two circles is your personal brand—how much people actually see of who you truly are. The overlap will look different based on the nature of your relationship with a particular person. Those people who know you and love you will have a lot of overlap. Someone meeting you for the first time or just scanning your LinkedIn profile will only see what you choose to show them.

It is worth it to analyze all the material and information you share with others publicly on social media, websites, and other channels you use to communicate professionally. You can use the following questions to complete this analysis:

- Does the content you typically share allow people to get to know you?

- If you are interested in career opportunities, does your messaging communicate what would make you uniquely qualified for the kinds of roles you are pursuing?
- If you were a recruiter or colleague, would you be impressed after viewing your profile or posts?

It's your responsibility to curate and control the narrative you present to the public, thus creating your personal brand.

A word of caution: As you work on your personal brand, don't try to be something you're not. Make sure what you share with others is authentic and truthful. Trust me—you don't want to be someone who creates a whole persona that doesn't even exist.

So be yourself, just the best version of yourself!

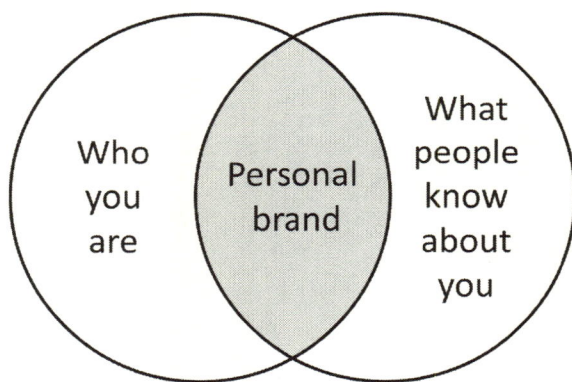

Figure 7.1: Personal branding diagram

WAYS TO UTILIZE A PERSONAL BRAND

If you utilize your personal brand well, your personal brand will help to increase your job opportunities, build your professional reputation, and grow your self-confidence.

As you consider what content to put out there for the world, it comes down to this—you want to clearly communicate what you do, the value you add, and how your skills and genius zones connect to what people or organizations need.

There are many ways you can work to build your personal brand, but here are a few key applications.

LinkedIn Profile and Resume

The content you create for your LinkedIn profile and resume is particularly important because these will, in many cases, be the first exposure a potential employer has to you. They provide a prime opportunity for you to showcase your personal brand.

Let's address LinkedIn specifically for a moment: LinkedIn is the world's largest professional networking platform. With more than 900 million users globally, LinkedIn provides unmatched chances to network with people from all areas of influence. If there is someone you want to meet or connect with, they are probably on LinkedIn.

It's also the place where people can find *you*—it may be the primary source of contact and exposure you have online, so your LinkedIn profile could be the hub of your entire personal brand.

As you create your LinkedIn profile and resume, it's important to balance the need to tell a story (like in your "About" section in LinkedIn or a "Personal Statement" in your resume), with using relevant keywords that employers are looking for.

So do your research. Learn what employers are looking for. Find a few job descriptions of roles that would interest you and look for patterns and themes.

Then, craft your profile and resume in a way that showcases that you are the type of person they are looking for! Bring forward the skills you have that are most relevant to the roles you want. Use the same language your potential employers use as you describe your talents, skills, and accomplishments.

When someone looks at your profile or resume, they should know exactly the type of work you do (or want to do), and how you can do it well.

Brenden, discussed earlier in the chapter, optimized his LinkedIn profile and was soon attracting attention from recruiters from the kind of innovative engineering and manufacturing companies he was interested in working for.

Networking Conversations

Your network is your net worth. —Porter Gale[71]

Regularly interacting with members of your existing network of connections and relationships and expanding your network, week after week, year after year, is an investment that will benefit you throughout your life. The sad thing is, many people don't work on networking until they really *need* their network. But if you're cultivating your network all the time, when you need it, you'll already have relationships with many people who can help.

Having career clarity will help you narrow in on what types of people you want to form relationships with. And having a clear personal brand will help the people you network with create a correct impression of you as you meet and talk with them.

One of the key things you'll want to do is create an "elevator pitch." An elevator pitch is a brief (short enough to be shared during an elevator ride) personal introduction that describes who you are and what you want to accomplish professionally. By crafting and memorizing an elevator pitch, you'll be prepared with a polished reply when someone says, "Tell me about yourself."

You'll have countless interactions like this as you meet people in your organization, attend networking events, conduct informational interviews, or just meet people as you move through your daily life. Wouldn't it be great to be ready for them? (I'll guide you through creating an elevator pitch in the "Take Intentional Action" section at the end of the chapter.)

When you have the opportunity to have longer interactions with a networking contact, whether at a networking meeting, a lunch appointment, or in an informational interview, you have an opportunity to weave in different themes from your personal brand that you want them to remember about you. This is where knowing your genius zones can really come in handy. You can also share some of your core values, the ways you do your best work, your primary interests, and areas where you want to grow.

If those you connect with resonate with what you tell them about yourself, it may open up opportunities you didn't expect, simply because you were proactive about sharing elements of your personal brand.

In Brenden's case, he started reaching out to friends, former classmates, and existing clients who he had relationships with who were doing things he thought were interesting. In those conversations, he was able to share things about himself—his expertise and his passions—that many of these people didn't know about him yet. His sharing piqued their interest and a few of his contacts made introductions to others he could talk to if he was interested in working at their organizations.

Job Interviews

A job interview is a prime opportunity to tell your story and control your narrative. In fact, many common interview questions ask you to do just that when they say, "Tell me about a time when...."

Telling stories in a job interview to illustrate your skills, past successes, and future possibilities at their company, makes the information you share about yourself much more interesting and memorable.

One great way to structure stories you tell in job interviews is using the widely-used STAR method. The STAR acronym breaks down like this:

- **Situation**—you need to set-up the situation in which the story takes place, whether at an event, during work, etc. Describe why the story you are telling matters.
- **Task**—clearly state what you wanted to achieve and your particular responsibility in the story (to distinguish it from the roles other people may have played).
- **Actions**—what are the steps you took in order for you to move through the situation. How did you work with others, what problems did you solve, and what was the process you used to accomplish the task given to you?
- **Results**—this is the outcome or the conclusion of the situation you are telling about. Share measurable results if possible. Wrap up the story in a way that demonstrates the positive outcome you achieved.

One of your main goals in an interview is for the interviewers to remember you. You want them to remember key elements of your personal brand. So when you're telling a story, incorporate elements of your top skills, genius zones, and core values.

For example, if the interviewer asks you about a time you experienced a conflict, first describe your views about how to best manage conflict, then share a specific example of how you did that. If they ask about a technical challenge you faced, briefly tell them about your philosophy and methods for solving problems, then back it up with a story of overcoming a big obstacle.

As you do this, it's important for you to share with potential employers how your skills, abilities, and genius zones align with their needs.

Here's another way to look at it: The fact that the interviewer is interviewing you for a job indicates they have a problem that needs to be solved, and they are looking for an employee to help solve it. The better you can tell your story and describe yourself in a way that presents *you* as the solution to their problem, the more likely you are to land the job.

When Brenden started getting job interviews, I helped him prepare by thinking through stories that would be useful for him to share. We did mock interviews so Brenden could practice answering possible interview questions with stories and weaving his personal brand into his stories.

Preparing in this way, combined with his high energy and charisma, helped Brenden land an excellent opportunity (with a significant pay raise) working for a growing space company. His new role fully utilizes his experience, passions, and skills, and gives him the opportunity to live in a new city that he's enjoying. Sounds pretty intentional to me!

FINDING MEANING IN YOUR STORY

In some way, suffering ceases to be suffering at the moment it finds a meaning, such as the meaning of a sacrifice. —Victor Frankl[72]

Personal branding isn't just about looking at your professional experiences. It's also about describing and communicating who you are as a human.

Everyone has a unique life story. You included. You may not be able to change your past experiences, but you can change your perspective about what your experiences *mean* to you, and perhaps find ways to communicate that meaning through your personal brand. The more you understand yourself and recognize there is a purpose and meaning to draw from every experience you've been through, the more you can wholeheartedly embrace the present and future.

Whether your past experiences paint a mostly happy or sad picture as viewed from the outside, you get to choose how you interpret your experiences. In doing so, you choose a path of empowerment rather than a path of victimhood as a pawn in the game of life.

Here are a few examples to illustrate reframing life experiences.

- Victor Frankl, was a Jewish psychologist who was imprisoned in four different Nazi concentration camps, including Auschwitz during World War II. His father, brother, mother, and wife died during this time as a result of their treatment by the Nazis. He endured horrendous experiences, and nearly all his freedoms were taken away. Even so, he chose to look for love, beauty, humor, and purpose in his life.

  Even before he was imprisoned in the camps, Frankl developed important psychological principles, which he called Logotherapy. He worked intently throughout his life to bring these principles to the world, and the desire to do so gave him the purpose and motivation he needed to survive the brutality of the concentration camps. This purpose also helped him realize he could use his experiences as an opportunity to learn about and expand his understanding of human behavior, thus helping him see a positive aspect to his terrible situation. Eventually, his experiences were recorded in his book, *Man's Search for Meaning*, which has widely become known as one of the most influential books ever written. I highly encourage you to read it.

- Ralph R. Teetor was a successful inventor, engineer, and philanthropist, best known for inventing cruise control, which most of us rely on in our vehicles and has helped make driving safer and more comfortable. He was also blind.

Despite his blindness, Teetor accomplished much. In fact, once a colleague asked him how much more he might have been able to achieve if he could see. Teetor, upon reflection, responded, "I probably couldn't have done as much. I can concentrate, and you can't." [73] In what most of us would consider a limitation, he found a strength.

- One of my clients, Nitin, was a recent graduate with a master's in mechanical engineering. However, there was a hiccup in his immigration paperwork, and following his graduation in May 2019, he couldn't legally work in the USA until February 2020. Finally, once he was legally able to work, he started getting interviews. But then the COVID-19 pandemic hit, and all his job prospects disappeared.

We met a few months later, when Nitin was working for minimum wage at a gas station, trying to make ends meet for his family. He was experiencing a lot of anxiety and felt depressed because of his situation. Through our work together, he was able to move past his feelings of worry and hopelessness and take intentional action that eventually landed him three job offers.

He received his first promotion after only eight months, and he is thriving! Being unemployed and underemployed was extremely difficult for him, but now he is grateful for the lessons he learned and how what he has learned helps him overcome the challenges that still come his way as he moves through his career and life.

How do you interpret your past experiences? What meaning are you creating from them? You get to decide what that meaning is and how you will communicate it to others.

TAKE INTENTIONAL ACTION

Create a Timeline

In order to extract meaning from your lived experiences, as well as dig deeper into understanding the "essence" of who you are, this exercise invites you to explore your past and create your own narrative. It's fun to see what comes of this.

1. Start by grabbing a blank sheet of paper or turning to an open page in your journal. Draw a horizontal line, a timeline, through the center of the page. The right end of the line represents the present day, and everything to the left of that point represents the past.

2. Reflect on your life thus far, and come up with five to seven core memories[74]—moments or experiences in your life that were defining in some way. These could be negative or positive experiences. Don't analyze them yet. Just jot them down and place them on the timeline.

3. For each core memory, ponder and write your answers to the following questions.

    ○ **What happened?** In asking this question, focus on the facts. Don't assign any meaning or purpose behind it yet. Just state the basic information. Age, people around, location, basic description of what transpired, etc.

    ○ **What did that moment mean to me?** This might be explaining why you chose this event as one of the core memories to include in this exercise in the first place. Was the meaning positive or negative? Why was the event important in your life? How did it change you?

    ○ **How do I want to think about this event from now on?** Ah, the moment of truth. You can look at every experience of your life and consider if the way you think about it is serving you, and if not, decide to assign a different meaning to it than you have in the past.

If you look back on one particular moment as terrible and negative, perhaps you can find lessons you learned from it that have strengthened you in ways that nothing else could. What did you gain from that experience?

4. After you've done a bit of analysis of each of the events using these questions, take a step back and look at it all. Are there any themes that emerge? How have you shifted the meaning of these events? How can you take more control of your story as it unfolds in front of you?

Doing this exercise with intention can bring healing and help you move forward with hope and purpose.

Create an Elevator Pitch

Having an elevator pitch you can use to break the ice in almost any conversation is a huge tool in your career transition toolbelt. Your "pitch" helps the person you are talking to understand who you are, what you do, why you do it, and what you want to accomplish professionally.

Memorize your elevator pitch and use it in networking conversations, job interviews, or in everyday casual conversation. You'll need to adjust your "pitch" to the context of the conversation at hand, but it's good to be prepared.

You'll want to create a few versions of your pitch to have at the ready when an appropriate situation to use one arises. Note that the examples and formats that follow are mostly designed for someone trying to make a career change, but you can adapt the principles to whatever you most want to communicate.

Five-Second Elevator Pitch

It's helpful to create a five-second elevator pitch to have on hand when you only have a brief moment in which to communicate who you are and what you do. Here are a few examples of how you can structure this version of your pitch:

- "I am a [insert role/profession] looking to make a transition into the [insert industry]."
- "I'm well-positioned for [insert role/function] due to my [background/experience; can insert # of years] working in the [industry/role]."
- "My big goal right now is to break into [insert industry] so I can use my [insert skills] to [insert the value you will bring]."

Thirty-Second Elevator Pitch

Your 30-second pitch is your longest, and probably most widely used, elevator pitch. It needs to include who you are, what you do, who you do it for, and what you want to accomplish professionally. Here is the basic structure:

I am a [insert role/profession] helping [say who you help] do [say how you help them]. My greatest passion is to [state your passion— this is your why]. What makes me different from most people is [list some of the biggest problems you've solved or some of your greatest accomplishments—why are you special?].

If it's a specific job you're after, you might finish by saying:

It's great to learn that [mention the opportunity] aligns perfectly with my career goals. Have you ever [finish with a question, such as "What are the most important traits/skills you are looking for in filling this role?"—keep the dialogue going!]?

If you're talking to someone who is not a potential employer, change the end of your pitch to: "I'm looking for opportunities that [mention the kind of opportunity you're looking for]. Do you know anyone who is looking to hire someone for that type of role?"

If you're not looking for a new job, but just expanding your network, here's one more way to finish: "I'm looking to grow my skills in [mention the areas in which you want to continue to expand your skills or expertise]. Do you have experience in that, or know someone who does?"

So if I were a mechanical engineer interviewing for an opportunity in the aerospace industry, I might say something like this:

> *I am a mechanical engineer helping aerospace companies optimize their designs for manufacturing success. One of my greatest passions is to see my designs get implemented and perform in critical environments. What's unique about what I do is that I have worked in design and technician roles, so I have the perspective of those who work on the manufacturing floor as I optimize mechanical designs. It's great to learn that this engineering role aligns perfectly with my career goals and aspirations! What are the most important traits/skills you are looking for in filling this role?*

As you brainstorm and create your own elevator pitch, here are a few questions to help you uncover the different elements you might want to share.

- Who are you and who do you help?
  - Sample answer: *I am a software engineer helping full-stack teams build great products that customers love.*
- Why are you passionate about your work?
  - Sample answer: *I love seeing my products being used by customers and clients and hearing how my products have transformed their lives in positive ways.*

- What makes you different?
  - Sample answer: *What makes me different from most people is that I started as a mechanical engineer and moved to software engineering, having built a range of systems that automate engineering processes and draw from massive databases.*
- How does the opportunity you're considering relate to your goals and aspirations?
  - Sample answer: *I'm thrilled to see that your company's vision of developing customer-centric software aligns with my desires to grow in a company that makes a positive contribution to the world.*

Don't be too formulaic about this, but you can use these ideas and examples as a guide.

Make Your Genius Zone Relevant

In the last chapter we went deep into the concept of genius zones and how to uncover them. But identifying your genius zones is only one part of the equation. Once you have identified them, they can be key pillars in how you communicate your personal brand.

To utilize one of your genius zones in a personal brand, you need to figure out how to describe it in a way that is relevant to those you want to attract. In particular, it's important to communicate how your genius zone creates value for others.

For example, one of my genius zones is my combination of broad engineering experiences, coaching/training skills, passion for helping others unlock their potential, and experience helping people to shift their mindsets. That's cool, but saying it like that might not mean much to potential clients.

To make it easier to understand, if I was talking to an individual interested in working with me, I would showcase how I help engineers intentionally upgrade their careers to create meaning and purpose in their lives by aligning their career and life. If I was sharing my expertise with an organization, I would focus on how I help technical professionals improve engagement and human connection to unlock productivity, creativity, and collaboration in impactful teams through innovative training and coaching programs.

So, how is your genius zone relevant to potential employers or clients? *That* is what you want to communicate as part of your personal brand.

Use The 5 Ps

Any time there is a job interview, a big meeting, a performance review, or some other opportunity to present your personal brand, you'll want to put your best foot forward. But these big moments don't just happen successfully on their own. You need to take action to make them successful.

How do you make sure you are ready to execute at a high level? What do you do in the days leading up to, and even the final moments before, an event to give yourself a good shot at doing a great job? Use the 5 Ps.

1. Prepare

Invest your time into preparing for the event. Collect relevant resources, get ready to answer tough questions with confidence, create any materials or visuals you want to share, do your research on who you'll be talking to, etc.

2. Pee

That's right, literally go to the bathroom! You need to relieve yourself so that later, needing to do so won't be a distraction for you. And while you're at it, make sure your hair and clothes are looking good, your zipper is up, you don't have any food between your teeth, your shoes are tied, etc. This step may be simple, but it can help you avoid a major embarrassment in a moment when you want to show up focused and confident.

3. Engage Your Physical Body

One of the best ways to get ready in the moments before you present yourself is to get your body active. You don't want to walk into a meeting or event feeling drowsy or lacking energy. Get your blood flowing! I like to do some push-ups or go for a walk, but you can get moving in any way you can think of that raises your energy without wearing you out.

4. Power Pose

One of the most widely watched TED Talks of all time is Amy Cuddy's where she talks about the power of body language. [75] Getting into a superhero pose with your hands on your hips or standing triumphantly with your hands raised in the air for a couple of minutes can decrease your stress level, and therefore increase your confidence going into a situation that is normally stressful for you. It may sound a bit weird, but I promise you it's worth trying. What do you have to lose?

5. Be Present

Finally, take a few deep breaths and be fully present in the moment. Release your worry about what preparation you could have done that you didn't—*that* time has passed. Don't worry about what the final outcome will be, as that moment hasn't arrived yet. Be as present as you can be with yourself and the people you're talking to, seeking to connect with them as humans. If you are fully present and make a connection, you're actually much more likely to stay relaxed and focused and to make the most of the experience.

To practice using the 5 Ps, in the next week or so, identify one event, meeting, presentation, interview, or conversation that feels particularly stressful and use the 5 Ps to prepare and execute it well.

CHAPTER 8

SKILLS THAT SUPPORT YOUR INTENTIONS

This chapter is a bit different from the rest of the chapters in this book—it is even more action oriented than the others. I discuss some of the most important aspects of life to focus on along your path of creating a life of intention.

In each section that follows, I've shared a few principles and provided you with an activity to help you strengthen your approach.

Ready? Let's go!

THE ART OF COMMUNICATION

> The art of communication is the language of leadership.
> —James Humes[76]

Are you an effective communicator?

Pause a moment before you answer.

If your instinct is to say "yes," let me ask another question—do those you work with every day believe you are an effective communicator?

It might be worth asking them. Even if you are pretty good at communicating, (if they are honest with you) they may share opportunities for you to improve your communication approach.

Communication is a *huge* aspect of your personal and professional success. So it's something you should never stop working on.

Simplify Communication

Everything should be made as simple as possible, but not simpler.
—Albert Einstein[77]

Engineers work on remarkably complex products, processes, and systems. Many go deep into learning and applying specialized technical knowledge. They need to learn new programming languages, understand complicated technical details, and put them together into something that is great for end users.

But here's the lesson—other than when you're communicating with your technical team, drop the technical jargon. Simplify your communication.

Often you need to communicate and (don't cringe) even "sell" ideas and initiatives. You need to explain why what you do is important and connect it to the needs of the team, department, and organization as a whole.

You will need to get buy-in from other teams that you need resources or assistance from.

No one will care to help if they don't understand what you are doing, and they won't understand if you don't simplify what you share with them.

Use analogies, metaphors, stories, physical demonstrations, simplified diagrams, or prototypes—whatever you need to use to help contextualize the information you share with others.

It may sound cliché, but it's a good idea to explain complex information in a way that even a fifth grader could understand your meaning, especially when speaking to non-technical personnel. I'm not suggesting you talk down to the other person or treat them like they are unintelligent, but rather, that you communicate simply and clearly so that the other person can apply their own expertise to the information you share with them. You will be able to accomplish much more if your communication is easy to understand.

Ask Great Questions

Without a good question, a good answer has no place to go.
—Clayton Christensen[78]

Let's take a step back and remember that communication isn't just one-way. Communication is about understanding *and* being understood.

That means you need to do more than just explain things yourself, but also ask great questions to understand others you are working with. This is true whether you're communicating with your team, your boss, a potential employer, or somebody from another team you're collaborating with.

What constitutes a great question will be different depending on the situation and who you are communicating with. Sometimes you need a definitive answer and need to ask a yes or no question. But often, the most powerful conversations are catalyzed by open-ended questions. These questions don't have one correct answer. They cause people to pause, think, consider, and process through their answers.

Once you get an answer, you can often go deeper in the conversation. Ask why. Ask clarifying questions. Ask to learn more about what they are sharing with you and how that connects to other ideas and challenges. Then, repeat back and summarize the information you've gathered to make sure you understand.

As an added bonus, when you ask great questions and learn more about what others care about, it helps you simplify and improve the information you share with them because of connections you've made between what you need and what they need! This process of asking great questions and thus improving your communication then reinforces your desire to simplify communication. Awesome, right?

Listen, *Really* Listen

When people talk, listen completely. Most people never listen.
—Ernest Hemingway[79]

Listening is hard for me. Maybe it's hard for you too.

A few years ago I received what is called a 360-Degree Review at work. Around 12 different people, including those I managed, peers I worked with, and my own leaders gave me anonymous feedback on my performance, my approach to teamwork and leadership, and my strengths and weaknesses.

The glaring theme from multiple sources was clear—I didn't listen well. I cut people off. I drove forward with my own ideas rather than listening and actually considering what others had to say. Some people felt plain ignored much of the time.

Receiving this feedback was painful. It wasn't what I had thought was true or wanted to be true, but getting that perspective became a turning point for me at work and at home (my wife confirmed that she felt I had trouble listening at home as well. Ugh...).

Working to improve my ability to listen and changing my negative tendencies is a journey, and one I feel I will never be done with. But through a lot of effort, I've made considerable progress. I had to have some honest conversations with my team to learn more about areas I needed to improve. We put the word "LISTEN" on our team whiteboard. I put "LISTEN" on a sticky note attached to my laptop. And I asked my team to hold me accountable as I tried to improve, so they started letting me know when I was cutting them off or ignoring their point of view. That experience changed me and improved our whole team dynamic.

There is power in listening. People want to feel heard. We have two ears and one mouth for a reason!

When people speak to us, we need to close off other distractions, look them in the eye, and be present with them.

There is a lot of talk about "active listening"—active listening is great, but we can go even further.

To build on the concept of active listening, Stephen R. Covey talked about "empathetic listening."[80] Empathetic listening means listening with your ears to what people are saying, but also listening with your eyes (to the speaker's body language), and heart (your feelings and the feelings of the person you're talking to). When you listen empathetically, you can often connect in a deeper way with the person you're speaking to, and thus accelerate the level of communication you have with each other.

It's amazing what can happen when you *really* listen to a person.

Communication Activity

For this activity, you'll practice empathetic listening. In order to do this activity, arrange a get together with one other person, and follow these directions.

1. Explain the purpose of the activity. The purpose is to practice empathetic listening, and thus gain a greater understanding of each other.

2. Have the other person describe a recent experience or respond to one of these prompts:
    o   Tell me about a time you were really scared, surprised, or happy.
    o   Share an achievement you are proud of.

       ○   Tell me about a formative experience from your childhood.

3. Have them share with you for three to five minutes.

As they share, your job is just to listen. Do not interrupt or ask questions during that time. Nothing. Make eye contact and try to understand the underlying emotions the person is experiencing as they share.

4. After they have shared with you, it's your turn—spend two to three minutes to reflect and share back what you heard and felt. You can use phrases such as, "It sounds like you felt (insert emotion or perceived sentiment) when (event happened). Is that right?" Verbally acknowledge the feeling surrounding what they shared. No need to analyze it further, just recognize the feelings you picked up on.

5. Next, switch sides and repeat the previous steps.

6. When done, debrief together. What did you learn about each other? What parts of this process were difficult for each of you? How will you try and listen differently from now on as a result of this activity?

This activity might sound awkward and uncomfortable to you. Great, do it anyway. It can be a transformational and relationship-enhancing experience.

BEING ACCOUNTABLE

Accountability is one of the most powerful motivators, but typically we think of accountability as someone else holding us accountable.

Even more powerful is when we choose to be accountable to ourselves.

Make an Appointment

A study by the American Society for Training and Development (ASTD; www.td.org) found that certain actions are more likely to lead to a person completing a goal.

- Just having the goal: likelihood of completing the goal is about 10%
- Consciously deciding you will accomplish your goal: 25%
- Setting a timeline for completing your goal: 40%
- Planning how you will do it: 50%
- Committing to another person that you will do it: 65%
- Setting up a meeting to discuss your progress with someone you've committed to: 95% likely to complete the goal![81]

That's astounding! Don't keep your goals and ideas to yourself. Share your goals with others, enlist an accountability partner, and then commit to them and follow up with them!

Create An Accountability Contract

I have a sweet tooth, and I'm often too lazy to exercise. That combination doesn't always bode well for my health goals.

One of the strategies that most helps me to stay motivated when I'm struggling is creating an accountability contract. Basically I create a list of things I will commit to. If I follow through, there will be some sort of reward. If I don't accomplish the objective, then there is a consequence.

I have a friend who has transformed his own health who has become my accountability partner. I'll commit to myself and my accountability partner about health actions I will take for a week, and if I don't keep them, then I owe him $10. That's not a lot of money, but I hate giving money to other people for stupid things, so it's rare that I don't follow through! (For me at this point in my life, avoiding a consequence is a stronger motivator than receiving a reward.) Sometimes this means I find myself doing a set of pushups late at night or skipping out on a dessert at an event, but I'm able to find that push because I'm accountable to someone else.

What kind of accountability contract can you create? Who can you set it up with?

Create a Culture of Accountability

If you work with a team at your job or in any other area of your life (think sports, community or faith groups, or even your family), you can create a culture that helps everyone be accountable.

In a successful team that has high levels of accountability, you will notice a few key elements:

- Everyone **seeks to understand** the goals of others. Of course this is true! How can you really know how best to help others if you don't know what their goals are? Taking responsibility to understand this is a critical step to being personally accountable.
- The actions individuals take **consider others' needs** so that they can gain greater collective success rather than focusing solely on their own accomplishments.
- There is a positive pressure from the team that drives accomplishment because they **expect great things** from each other.
- Team members **frequently check in** with each other to assess how their own actions are impacting the others.

Does your team incorporate these elements? If not, invite your team members to discuss how your team can incorporate accountability to be more successful.

Accountability Activity

In addition to taking actions I've already discussed in this section, including enlisting an accountability partner, creating an accountability contract, and making an appointment with your accountability partner, here's an activity for you to do to help improve accountability in a team setting.

Have you ever, at the end of a meeting, said to yourself, "Okay, we talked about things, but I have *no* idea what I'm supposed to do now...."

If you answered yes, that's likely because there weren't clear decisions or assignments made, which leads to minimal follow-through and consideration afterwards. Here's how you can fix that for your next meeting, assuming you are in a position of leadership or at least can suggest this change in an appropriate way.

- At the end of the meeting, review key decisions and what needs to be communicated outside of the meeting. **Make assignments for each action item**!
- If a decision cannot be made immediately (due to a lack of information, insufficient time in the meeting to come to a decision, etc.), **set a clear deadline** for when a decision *will* be made by.
- **Publicize the team's goals**, who is responsible for deliverables, and what the expectations for each team member are.
- Individually, identify key people who are dependent on your work. **Regularly check in** with them on how well you are doing in helping them.
- **Shift reward structures** and incentives away from individual performance and toward team achievement.

If you are not in a position of leadership, consider having a private conversation with the person who leads your meetings about how you can help make your organization more successful by incorporating these ideas into meetings.

GIVING AND RECEIVING FEEDBACK

Giving and receiving feedback is sort of an art and a science. It's such an important precursor to growth, and yet most people dread it. Perhaps you've been lambasted and degraded by others, or perhaps in the past you gave someone feedback in an attempt to help them, but they took offense. Fear of the potentially negative outcomes of giving and receiving feedback is often the driving force behind feedback avoidance.

Don't let this be the case with you. Be courageous and allow feedback to empower you and those you live and work with.

Giving Feedback

The gift of truth excels all other gifts. —Gautama Buddha[82]

There will be moments in your career when you need to tell a hard truth. Though it's likely to be an uncomfortable experience, and you're unsure how your feedback will be received, it is worth it.

To distinguish this kind of thinking, you must come into these sometimes difficult conversations with a mindset of genuinely wanting to help the other person and from a place of caring and concern. If you genuinely care about a person and want to help them, but you don't share what you feel is important, then you are withholding an opportunity for crucial development for this individual. Of course, I'm not suggesting you need to be in everyone else's business and point out all the shortcomings of every member of your team. I'm referring to situations when you genuinely feel it's important for you to share feedback with someone.

If a person doesn't have the truth, then they can't operate from reality. All of us are imperfect people that have the capacity to grow as individuals, as part of society, and as part of a team. Yet in order to do that, we need to know where we need to focus to enhance our skills and improve who we are as people.

These difficult conversations are necessary and have far-reaching effects on everyone involved. The truth could be hard to share or to digest, but if you can approach having these conversations in a gentle, positive, supportive way, the receiver is more likely to have an open mind and take it as a learning opportunity. They can take your feedback as the gift of truth that it is.

Receiving Feedback

If you are on the receiving end of difficult feedback, take a deep breath and truly listen. It doesn't matter if this feedback is coming from someone who is your leader or even someone who you lead, it's important for you to handle it with humility and openness.

Here are a few ways you can do that.

- **Consider feedback a gift.** As I encouraged before regarding giving feedback, remind yourself that the feedback you're receiving is a gift, and treat it as such. Show appreciation and thank the person for being honest and sharing the feedback with you. Indicate that you will take it seriously. You want to validate them for having the courage to share this with you.
- **Get curious.** Rather than just listening and accepting exactly what is shared with you, get curious and ask follow-up questions. Dive deeper into the root of the issue and what really needs to change.
- **Seek for multiple perspectives.** What you don't want to do here is go seeking allies who will just placate you and tell you what you want to hear. When there is something that needs to change, there are usually multiple people involved. Seek out two to three other people who are close to the situation that might be able to confirm and add

more depth to your understanding of the feedback you've received.

- **Apply the growth mindset.** Remember, when you have a growth mindset, you see challenges as an opportunity to learn and grow. Maintaining a growth mindset will help you take feedback as a constructive opportunity rather than a personal slight, and can even help you encourage others to give you feedback.

- **Summarize what you hear.** Once you feel like you have heard what has been shared with you, summarizing the feedback can help the other person feel heard. This ensures that everyone is on the same page, and can help you come to a place of clarity about what specific actions you need to take after the conversation. Feedback is no good if it's lost in translation, so summarize it and validate your understanding.

You don't have to agree with or 100% accept every piece of feedback you receive. Different people might think you need to change in different ways, and there's no way to please everybody. The key is to create intentions for who you want to be and how you want to operate in the world, then use feedback opportunities as a chance to help assess how you are doing at living congruently with your intentions.

Curating Psychological Safety

For knowledge work to flourish, the workplace must be one where people feel able to share their knowledge! This means sharing concerns, questions, mistakes, and half-formed ideas.
—Dr. Amy Edmondson[83]

One of the most important factors to delivering effective feedback is creating an environment that is open and safe for the giver and receiver. If people feel they will be retaliated against if they share feedback, or have had people share feedback in ways that have been aggressive or disrespectful, the culture is one that discourages giving feedback, at the detriment of the organization.

Cultivating psychological safety may be the answer. In fact, a few years ago, Google did a large study of the most effective teams at the company and found psychological safety to be the #1 factor of team success. The researchers defined psychological safety as "an individual's perception of the consequences of taking an interpersonal risk."[84]

Teams feel psychologically safe when:

- Team members feel connected to each other.
- People feel heard, seen, respected, and listened to.
- If a team member tries something and it doesn't go well, they know they will be supported because they feel like people on their team care about them.
- They feel like they are able to ask for help from others.
- They feel like their unique talents and skills are utilized and valued.
- They know they can take risks.
- People feel like they can bring up tough issues.

Yep, these are all things that invite, encourage, and make feedback a positive experience for all.

Feedback Activity

The best way to practice giving and receiving feedback is just doing it. Here's an activity to get you started on inviting and being open to receiving feedback.

Try this:

1. Identify a person with whom you would like to improve your work or personal relationship.

2. Brainstorm three to five changes you can make to be more helpful to them. For example, you could express appreciation when they follow through on an important assignment. Or you could make an effort to be present when they are sharing important information with you.

3. Invite this person to spend 20 to 30 minutes with you. Let them know you will be asking for their honest feedback.

4. When you meet with them, start the conversation by thanking them for taking the time to talk with you. Then share the areas you believe you can improve to be more helpful to them. Ask if they agree with these items and if they would like to clarify or add anything.

5. Keep asking clarifying questions until you feel you truly understand their perspective.

6. Again, thank them for being willing to share with you, and let them know you'll be making an effort to improve.

If you make it clear to others that you are humble, open to feedback, and serious about improving, you will create an environment where others feel safe openly sharing honest feedback with you.

STRENGTHENING RELATIONSHIPS

One of the longest-run studies ever conducted is the Harvard Study of Adult Development. It started by tracking the health of 268 Harvard sophomores (all men) starting in 1938, and has continued since then, following the men and even their offspring.[85]

Here is the surprising finding: The people with the best health were those with the most satisfying relationships. Close relationships, more than money or fame, are what keep people happy throughout their lives.

"When we gathered together everything we knew about them at age 50, it wasn't their middle-age cholesterol levels that predicted how they were going to grow old," said Robert Waldinger, the study's director, in a popular TED Talk. "It was how satisfied they were in their relationships. The people who were the most satisfied in their relationships at age 50 were the healthiest at age 80."[86]

In May 2023, the US Surgeon General put out an advisory stating that loneliness and isolation are causing detrimental health effects at an epidemic level. Here are just a few of the findings:

- Isolation is costly. Social isolation among older adults accounts for an estimated $6.7 billion increase in Medicare spending each year, mostly in hospital and nursing facilities. Also, work absenteeism attributed to loneliness is estimated to cost employers $154 billion each year.
- Loneliness reduces achievement at work and school.
- Social isolation increases the risk of premature death by ~29%. It's estimated to have the same effect on mortality rate as smoking 15 cigarettes per day.
- Isolation increases the risk of heart disease, diabetes, and obesity, as well as mental health problems such as anxiety, depression, and dementia.[87]

Relationships matter. I'm increasingly convinced that it is impossible to have a meaningful, happy, and healthy life without great relationships. Certainly this includes our relationships with our spouse or partner, siblings, children, and other family members, but also close friends, collaborators, neighbors, and more.

Here are a few ideas to build meaningful and lasting relationships that will enrich your life.

Be Transformational, Not Transactional

Always remember that you are trying to build relationships that are *transformational* not *transactional*—both you *and* the other person should benefit from the relationship! Together, you collectively get more out of your interactions than you ever would if you never met. That's the spirit of any strong relationship.

Here are a few ways to ensure your relationships are transformational, not transactional.

- **Be a giver, not just a taker.** Don't show up to meetings or interactions with your primary goal being to get something out of the other person. Always be looking for ways to contribute to the relationship. You have the opportunity to be a positive contributor to every relationship you encounter by serving the other person. Learn what they care about. Help them connect with a person they would benefit from connecting with or an idea that would be useful to them. Offer to help them on a personal or professional project. Or, just be there in a time of need. Certainly you can ask for help and favors too, but make sure it goes both ways.

- **Focus on the other person.** No one likes it when they can tell that the person they are interacting with is only focused on themself. When you are in conversations with others (or even writing emails), how often do you find yourself saying "I" and "me"? If it's a high percentage of most of your interactions, there is probably an adjustment to be made. Instead, focus on them. Turn your attention towards the other person and what they care about, are thinking about, and are trying to work on these days. This shows your desire to know and invest in them.

- **Ask great questions.** The power of inquiry is amazing. When you ask great questions, you show your curiosity and interest in the other person. Go deep and see what you can discover about them. Listen intently. Few things can show you care in a relationship than asking and listening.

Be Known

Have you ever found yourself feeling lonely in a room full of people? I know I have.

This usually happens when I feel like no one there truly knows me or cares about me. The conversations feel surface-level and superficial. I can only ask and answer, "So, what do you do?" so many times....

Staying surface-level never deepens relationships. I have often felt insecure and afraid that if people knew the "true" me, they would dislike me. So, I put on my best facade and tried to play the part I thought everyone wanted from me.

But as I've come to find out, when I'm willing to be vulnerable and let others get to know me, I feel more connected. I feel seen. I feel known.

This doesn't mean I need to air out my entire life story, my regrets, and my most personal feelings to everyone I meet. I save that for those I am closest with and have high levels of trust for. Then, when they accept me despite my many imperfections, I feel a greater sense of emotional security.

Being authentic in these situations allows people I trust to see the real me, and perhaps gives them the space to be more authentic and vulnerable as well.

Allow yourself to be known. In doing so, you'll invite more acceptance and connection with others in your life.

Disagree Better

Being able to disagree effectively is a really useful skill. Disagreeing effectively is more than just being nice. It's about disagreeing in ways that help us find solutions and even improve our relationships.

One indication that in our society, there is still work to do in this area is that each Thanksgiving, the internet goes wild with tips and advice on how to avoid fist fights at your family's Thanksgiving gathering.

Sure, there are divisions because of political and sports team loyalties, but it's more than that. We live in what author Arthur C. Brooks calls a "culture of contempt"—a culture of seeing people who disagree with us as not merely incorrect, but as worthless and defective. [88] John M. Gottman, a social psychologist and relationship expert who has studied thousands of married couples, has found that couples who often show contempt, including sarcasm, sneering, hostile humor, and eye-rolling, are much more likely to divorce.[89]

But we can disagree without contempt; disagreement doesn't have to be destructive. In fact, when we disagree effectively, conflict can actually be quite healthy. Disagreement can help us refine our opinions, elevate our goals, and work towards improvement. We don't need to disagree less, but to disagree better.

How do we do that? With *love*.

When we face disagreements with love, we can still see the good in others. We challenge the idea, not the person. We focus on *what* is right, not *who* is right. We might even go so far as to work to understand the other person's perspective so well that if we needed to, we could argue their point for them!

In the end, it's okay to agree to disagree. Just do it with love and respect, not contempt or resentment.

Do you disagree well?

Relationships Activity

If there is a personal or professional relationship you would like to deepen, I encourage you to find an opportunity to take some time where the only agenda item is to deepen your understanding of them as a person.

Come prepared with questions you want to ask and things you want to know about them. Include questions that encourage you to understand them as a whole person, not just in your working relationship (although that's important too).

Professional questions to consider:

- What are your main goals/objectives/priorities at work?
- What do you like about your job?
- What are the biggest challenges you're facing at work?
- Who are the key stakeholders that rely on what you do?
- What's your educational/professional background?
- Where do you see your career headed?

Personal questions to consider:

- Tell me about your family.
- What's on your bucket list?
- What's your favorite place in the world?
- Where do you find happiness/joy?
- What's your favorite dinosaur?
- What are three things you're good at?
- When you're not at work, where do you spend your time?
- If you could go anywhere, where would it be and why?

This is about asking questions and listening to get to know the other person better. Take notes on what you learn. Perhaps what you learn about them will help you identify a way you can assist them in some way. Thank them for their time, validate what they shared, and continue to find opportunities to deepen the relationship in the future.

CONCLUSION

WHAT'S IT ALL FOR?

Making changes in our lives and career can be scary. But when we know our *why*, when we figure out what we want to do and what it's all for, we can muster up the courage to act.

"I want to help engineers be better humans."

This was just one of the answers I gave when asked what I wanted to do as I left my corporate job and started my own training and coaching practice.

"I want to make money helping people," was another answer I gave.

I took a leap and didn't really know what I was getting myself into. (Maybe I still don't.) I had two kids (with plans to add a third soon) and a mortgage, and my wife was a full-time homemaker. I had no idea how I was going to make money to take care of them, but I had a strong intention with a lot of energy behind it.

With a growing family, I wanted to primarily work from home and be able to work on the road when we were traveling. I wanted to have flexibility in my schedule to take vacations, go to my kids' extracurricular activities, and welcome my kids when they arrived home from school. I wanted to create space to read, journal, reflect, and transform my personal character defects.

Now, as I write this, it's been four years since my last day working at the office, and sometimes I have to pinch myself. To an extent, I'm "living the dream." I love what I do. I get paid to help people (primarily engineers and tech professionals), and I've had many people tell me that they feel like they are becoming better humans and are happier in their lives because of our work together. And now, I'm an author, another long-held dream, inspired by many mentors of mine.

I'm not done growing, transforming, and impacting others, and neither are you!

This book lays out an approach to help you figure out what you want to do, what you're amazing at, and how you can create and be successful living the life you want to live. The interesting thing about success is that we all define it differently at different stages in our lives. What looks like success to me now is different than it was five years ago, and will be different five years from now. We change and transform. Additionally, we are all unique. Every client I talk to has different goals and dreams. That's excellent! These differences allow us to each bring our unique contributions that make the world an interesting place to live!

As you think about what you want for your future, I invite you to ponder the question in the title of the book, *How Will You Measure Your Life?*[90] In other words, what is the key measurement of a life well-lived? What is *your* answer? I imagine your key measurement isn't the amount of money in your bank account or your job title, but something else that truly matters to you. What is that thing?

The book's author, Clayton Christensen, was an influential author of many books, Harvard Business School professor, and consultant. He wrote *How Will You Measure Your Life* as he neared the end of his life, as I imagine he was contemplating the meaning of it all. It's a question I ponder often, and I invite you to do the same.

I truly believe that happiness, satisfaction, and joy come from building and nurturing strong connections with other people, contributing to causes we care about, doing good in the world, and connecting with God, a Higher Power, or something that gives us a sense of meaning outside ourselves. As an added bonus, when we make these strong connections, they often propel us to obtain many of the external results—money, jobs, travel, etc.—we are looking for. And it's not just the results themselves, but what they mean for us and those we are working to impact for the better.

If you recall the model from the beginning of Chapter 1 (figure 1.1), being intentional and creating purposeful growth in our lives is simply a pathway that leads us to having an impact as we engage in our life's work. It's a process and a transformation we all have the opportunity to make. I hope this book, one small part of engaging in my life's work, has impacted you and propelled you to live intentionally and make a positive impact in the world.

I hope that as you go forward, you will find meaning in your career and life and that you will accept some simple truths:

1. You have value.
2. You can always change and improve.
3. You deserve joy.

We've covered a lot of ground in this book. If you haven't already, it's time to take action on all that you've learned. Answer the call to live a life of meaning. Start where you are and go from there. Design your life with intention. Cultivate your mindset. Get clarity and discover your genius zones. Learn to communicate your best self through building a personal brand. Develop skills that will strengthen your resolve to live intentionally.

Throughout this book you've been asked to ponder dozens of questions about yourself and what is important to you. Go back through them and review your answers. And continue using a journal every day to create your intentions and live the life you desire.

Being intentional is a choice that only you can make—no one else gets to determine the life you live. Now is the time! Choose to be an Intentional Engineer, and create a life and career that you love!

ADDITIONAL FREE RESOURCES AND MORE

Check out the free companion resources I've put together to help you implement the principles from this book.

Website: www.jeff-perry.com
Blog: www.jeff-perry.com/blog
Free resources: www.jeff-perry.com/resources
Free digital version of *The Intentional Engineer Workbook*:
www.TheIntentionalEngineer.com/workbook

SOCIAL MEDIA

LinkedIn: www.linkedin.com/in/jeffcperry
YouTube: www.youtube.com/@jeff-perry

CONTACT, SPEAKING, OR OTHER INQUIRIES

Email: jeff@jeff-perry.com

ACKNOWLEDGEMENTS

If I have seen further, it is by standing on the shoulders of giants. —Isaac Newton[91]

I've always wanted to write a book, but was never sure when and how it would happen. I suppose I waffled for many years on trying to get my own clarity and upgrade my mindset to the point where I felt I really had something to say.

But it's not something I did alone. I have many people to thank.

A huge thank you to Callum McKirdy. It was in a conversation with you that I found the inspiration and spark to commit to writing this book, and I never looked back.

Thanks to the entire Thought Leaders community. Being a part of such a supportive group has helped me hone my messaging and receive support as I create my own intentional future and life by design. I'm especially grateful to those on our "book writing quest" including Danette Fenton-Menzies, Christopher Miller, Lisa Gerber, Anna Glynn, Hannah Brown, Tom Loest, Jet Xavier, Susanne Le Boutillier, James Hudson, and others.

Thanks to Josh Steimle. Your resources on LinkedIn and thought leadership inspired me early in my entrepreneurial journey, and your *Published Author Workbook* gave me a structure and a path to turn my idea for this book into a finished product.

To Brynn Steimle, whose patient editing and marvelous support turned my original manuscript into something that people might actually enjoy reading. You helped provide the fuel to get me over the finish line.

To my amazing clients over the years. Your stories are all over this book, and having a front-row seat to your success and growth is one of the greatest honors of my life. Thank you for trusting me to support you, and I look forward to seeing more intentional success in your future.

To my parents, Rob and Jan, for helping me grow up with all the necessities of life and so much more, including love, faith, and people who always believed in me.

To my four amazing kids. Your boundless energy and enthusiasm inspires me every day. Being your father is a huge part of the intentional future I am creating.

And to my dear wife, Robin. It's impossible to put into words how much your love and support mean to me. You were the first reader and editor of the book, as well as the cover artist. The fact that you thought this book was worth sharing means I don't care if anyone else reads it. Looking forward to intentionally living with you for eternity.

To my Heavenly Father and Jesus Christ. Thank you for blessing me with a beautiful life to learn and grow in. With your help, I can continue to change and become who I am meant to be. Thank you for guiding me when I need it, and letting me choose for myself when I need that as well, even when I don't like it. One step is enough for me. Thank you for loving me as your Son, and creating an eternal possibility for all of us.

NOTES

1 Richie Norton. Quoted by Amy Sargent-Kossoff, "New Year's Resolution: I Will Be Intentional." Institute for Social and Emotional Intelligence. January 2, 2023. https://www.isei.com/blog/new-year-s-resolution-i-will-be-intentional.

2 Gallup. "State of the Global Workplace: 2023 Report." https://www.gallup.com/workplace/349484/state-of-the-global-workplace.aspx.

3 Selcuk Eren, Allan Schweyer, and Malala Lin. "Job Satisfaction 2023: US Worker Satisfaction Continues to Increase." The Conference Board. https://www.conference-board.org/research/job-satisfaction/US-worker-satisfaction-continues-to-increase.

4 Thomas Tredgold. December 29, 1827. Quoted by F.R. Hutton, "The Field of the Mechanical Engineer." *The Engineering Digest*, 3, 10, (1908). See https://todayinsci.com/T/Tredgold_Thomas/TredgoldThomas-Quotations.htm.

5 National Society of Professional Engineers. "Engineers' Creed (2021)." https://www.nspe.org/resources/ethics/code-ethics/engineers-creed.

6 J.R.R. Tolkien. *The Hobbit*. New York: Houghton Mifflin Company, 1996, p. 7.

7 Joseph Campbell. *The Hero with a Thousand Faces*, third edition. Novato, CA: New World Publishing, 2008. (First published in 1949. See the Joseph Campbell Foundation, https://jcf.org.)

8 Wayne Dyer. Quoted by Mirela Xhota, "Our Intention Creates Our Reality ~ Dr Wayne Dyer." March 22, 2018. https://www.linkedin.com/pulse/our-intention-creates-reality-dr-wayne-dyer-mirela-xhota.

9 Abraham Lincoln. Quoted by John Kennedy, "Sen. Kennedy (R-La.) Issues Statement on Sen. John McCain." August 25, 2018. https://www.kennedy.senate.gov/public/2018/8/sen-kennedy-r-la-issues-statement-on-sen-john-mccain.

10 Brad Wilcox. "Five Easy Ways to Make School Hard and Five Hard Ways to Make School Easy." *New Era*. April 2009. https://www.churchofjesuschrist.org/study/new-era/2009/04/five-easy-ways-to-make-school-hard-and-five-hard-ways-to-make-school-easy.

11 Thomas Edison. Quoted in Benjamin Hardy, "How This 10-Minute Routine Will Increase Your Creativity." *Inc*. April 20, 2016. https://www.inc.com/benjamin-p-hardy/this-10-minute-routine-before-and-after-sleep-will-increase-your-creativity-and-.html.

12 Squire Bill Widener. Quoted by Theodore Roosevelt, *Theodore Roosevelt: An Autobiography*. New York: Charles Scribner's Sons, 1913, p. 337.

13 Jack Kornfield. *The Wise Heart: A Guide to the Universal Teachings of Buddhist Psychology*. New York: Bantam Books, 2009.

14 Stephen R. Covey. *The 7 Habits of Highly Effective People: Powerful Lessons in Personal Change*. New York: Simon & Schuster, 2004, pp. 99–100.

15 See Benjamin Hardy, "How To Get Precisely What You Want." *Medium*. January 1, 2019. https://medium.com/thrive-global/how-to-get-precisely-what-you-want-9cd39134538b.

16 Tony Robbins. "Where Focus Goes, Energy Flows." March 22, 2016. https://www.youtube.com/watch?v=b8jS86OtgLA.

17 Virginia Satir. Quoted in George S. Everly, Jr., "Time to Create the Life You Want." *Psychology Today*. January 1, 2022. https://www.psychologytoday.com/us/blog/when-disaster-strikes-inside-disaster-psychology/202201/time-create-the-life-you-want.

18 Brené Brown. *Atlas of the Heart*. New York: Random House, 2021, p. 84.

19 Steven Pressfield. *The War of Art: Break Through the Blocks and Win Your Inner Creative Battles*. New York: Black Irish Entertainment, 2002, p. 40.

20 C. Northcote Parkinson. Essay from November 19, 1955. As printed in *The Economist*. https://www.economist.com/news/1955/11/19/parkinsons-law.

21 Juran. "Pareto Principle (80/20 Rule) & Pareto Analysis Guide. March 12, 2019." https://www.juran.com/blog/a-guide-to-the-pareto-principle-80-20-rule-pareto-analysis.

22 Bill Burnett and Dave Evans. *Designing Your Life: How to Build a Well-Lived, Joyful Life*. New York: Knopf, 2016.

23 Rhianna Taniguchi. "How to Prototype Life-Designs." Designing Your Life. April 28, 2020. https://designingyour.life/how-to-prototype-life-designs.

24 Harvey Mackay. "Values Determine Who We Are." May 1, 2017. *The Business Journals.* https://www.bizjournals.com/bizjournals/how-to/growth-strategies/2017/05/values-determine-who-we-are.html.

25 John C. Maxwell. *Today Matters: 12 Daily Practices to Guarantee Tomorrow's Success.* New York: Center Street, 2008, p. 261.

26 Covey, *The 7 Habits of Highly Effective People,* p. 298.

27 *Oxford English Dictionary.* Accessed September 8, 2023. https://www.oed.com/search/dictionary/?scope=Entries&q=congruence.

28 Mihaly Csikszentmihalyi. *Flow: The Psychology of Optimal Experience.* New York: HarperPerennial, 1990, p. 224.

29 Ken Blanchard. Quoted by Paula Black, "Are You Committed or Are You Just Interested?" *Forbes.* April 16, 2021. https://www.forbes.com/sites/forbescoachescouncil/2021/04/16/are-you-committed-or-are-you-just-interested.

30 Steve Maraboli. *Life, the Truth, and Being Free.* Port Washington, NY: A Better Today Publishing, 2009. https://www.goodreads.com/quotes/318961-once-your-mindset-changes-everything-on-the-outside-will-change.

31 See https://arbinger.com.

32 The Arbinger Institute. *The Outward Mindset: Seeing Beyond Ourselves.* Oakland: Berret-Koehler, 2016, p. 12.

33 The Arbinger Institute, *The Outward Mindset,* p. 17.

34 The Arbinger Institute, *The Outward Mindset,* p. 18.

35 The Arbinger Institute, *The Outward Mindset,* p. 19.

36 The Arbinger Institute, *The Outward Mindset,* p. 19.

37 Carol Dweck. *Mindset: The New Psychology of Success.* New York: Random House, 2006, p. 29.

38 Dweck, *Mindset.*

39 Carol Dweck. "The Power of Believing That You Can Improve." TEDx Talk. November 2014. https://www.ted.com/talks/carol_dweck_the_power_of_believing_that_you_can_improve.

40 The Arbinger Institute, *The Outward Mindset,* p. 32.

41 The Arbinger Institute, *The Outward Mindset,* p. 12.

42 Peter Economy. "How the Platinum Rule Trumps the Golden Rule Every Time." *Inc.* March 17, 2016. https://www.inc.com/peter-economy/how-the-platinum-rule-trumps-the-golden-rule-every-time.html.

43 The Four Stages of Competence model has been used in many contexts, but was first used in the textbook *Management of*

Training Programs by three management professors at New York University (see the following note).

[44] Adapted from Frank Anthony De Phillips, William M. Berliner, and James J. Cribbin. "Meaning of Learning and Knowledge." *Management of Training Programs*. Homewood, IL: Richard D. Irwin, Inc., 1960, p. 69.

[45] Ryan Gottfredson. *Success Mindsets: Your Keys to Unlocking Greater Success in Your Life, Work & Leadership*. New York: Morgan James Publishing, 2020.

[46] Robert Kegan and Lisa Laskow Lahey. *Immunity to Change: How to Overcome it and Unlock the Potential in Yourself and Your Organization*. Boston: Harvard Business Press, 2009, p. 47.

[47] Siegrid Löwel and Wolf Singer. "Selection of Intrinsic Horizontal Connections in the Visual Cortex by Correlated Neuronal Activity." *Science*, 255, 5041 (January 1992). https://www.science.org/doi/10.1126/science.1372754.

[48] Brené Brown. *The Gifts of Imperfection: Let Go of Who You Think You're Supposed to Be and Embrace Who You Are*. Center City, MN: Hazelden, 2010, p. 75.

[49] Marilee Adams. *Change Your Questions, Change Your Life*. San Francisco: Berrett-Koehler Publishers, 2019.

[50] Carl Jung. Edited by Gerhard Adler and Aniela Jaffé. *Letters Volume I*. New York: Princeton University Press, 1992, p. 33.

[51] Lewis Carroll. *Alice's Adventures in Wonderland*. Chicago: VolumeOne Publishing, 1998 edition, pp. 89–90.

[52] James Clear. https://jamesclear.com/quotes/most-people-think-they-lack-motivation-when-they-really-lack-clarity.

[53] Tony Robbins. https://www.tonyrobbins.com/tony-robbins-quotes/inspirational-quotes.

[54] Philip McKernan. "In the Absence of Clarity, Take Action." *Medium*. March 28, 2020. https://philipmckernan.medium.com/in-the-absence-of-clarity-take-action-25c0c5359b96.

[55] Jeffrey R. Holland. "Wrong Roads." YouTube video. November 5, 2013. https://www.youtube.com/watch?v=yNQC-_srxH8.

[56] Tony Robbins. https://www.linkedin.com/posts/officialtonyrobbins_stay-committed-to-your-decisions-but-stay-activity-6309044479009660928-bHFp. Quoted in Emily Conklin, "9 Powerful Tony Robbins Quotes That Will Redefine Your Quest for Success." *Entrepreneur*. September 22, 2017. https://www.entrepreneur.com/leadership/9-powerful-tony-robbins-quotes-that-will-redefine-your/300297.

[57] Friedrich Nietzsche. *Twilight of the Idols*. Indianapolis: Hackett Publishing Company, 1997, p. 6.

58 See https://asq.org/quality-resources/five-whys.

59 Covey, *The 7 Habits of Highly Effective People*, p. 288.

60 Richard G. Scott. "Finding the Way Back." *General Conference Report. April 1990.* https://www.churchofjesuschrist.org/study/general-conference/1990/04/finding-the-way-back.

61 Gay Hendricks. *The Big Leap: Conquer Your Hidden Fear and Take Life to the Next Level.* New York: HarperOne, 2009.

62 Csikszentmihalyi, *Flow*.

63 Laura Garnett. "Define your zone of genius: Laura Garnett at TEDxMillRiver." November 10, 2012. TEDx Talk. https://www.youtube.com/watch?v=gQ7_r2oWlrw.

64 Hendricks, *The Big Leap*, p 144.

65 Laura Garnett. "Define your zone of genius: Laura Garnett at TEDxMillRiver." November 10, 2012. TEDx Talk. https://www.youtube.com/watch?v=gQ7_r2oWlrw.

66 Csikszentmihalyi, *Flow*, p. 42.

67 Mihaly Csikszentmihalyi. "Flow, the Secret to Happiness." TED Talk. February 2004. https://www.ted.com/talks/mihaly_csikszentmihalyi_flow_the_secret_to_happiness.

68 Adapted from Csikszentmihalyi, *Flow*, p. 74.

69 Adapted from Laura Garnett, TEDx Talk.

70 See https://www.youtube.com/watch?v=Spq9_TIsmeY&ab_channel=LIDA360%2FlidaCitroen.

71 Porter Gale. *Your Network Is Net Worth.* New York: Atria Books, 2013.

72 Victor Frankl. *Man's Search for Meaning.* New York: Washington Square Press, 1984, p. 135.

73 Blind Logic. "How Ralph R. Teetor Overcame Adversity Despite Being Blind." October 14, 2022. https://www.blindlogicproductions.com/how-ralph-r-teetor-overcame-adversity-despite-being-blind.

74 "Inside Out." Movie directed by Pete Doctor (Pixar, 2015). See Peachey Counseling, "What are Core Memories and Do They Matter?" https://www.peacheycounselling.ca/blog/2021/q-and-a-about-core-memories.

75 Amy Cuddy. "Your Body Language May Shape Who You Are." TEDGlobal Video. June 2012. https://www.ted.com/talks/amy_cuddy_your_body_language_may_shape_who_you_are.

76 James Humes. "The Art of Communication Is The Language of Leadership." *Fresh Business Thinking.* March 27, 2008.

http://www.freshbusinessthinking.com/the-art-of-communication-is-the-language-of-leadership.

[77] Albert Einstein. Quoted by Alice Calaprice, *The Expanded Quotable Einstein*. Princeton, NJ: Princeton University Press, 2000, p. 314.

[78] Michael Bungay Stanier. *The Coaching Habit: Say Less, Ask More & Change the Way You Lead Forever*. Toronto: Box of Crayons Press, 2016, p. 88.

[79] Ernest Hemingway. Quoted in A.E. Hotchner, *The Good Life According to Hemingway*. New York: Ecco, 2008, p. 113.

[80] Covey, *The 7 Habits of Highly Effective People*, pp. 239–243.

[81] Association for Talent Development. As referenced by Angela Hayes, "Being Held Accountable for Your Goals." Colorado State University Alumline. October 10, 2019. https://alumline.source.colostate.edu/being-held-accountable-for-your-goals.

[82] Narada Thera. *The Dhammapada: The Gift of Truth Excels all Other Gifts 1940*. Whitefish, MT: Kessinger, 2004.

[83] Amy C. Edmondson. *The Fearless Organization: Creating Psychological Safety in the Workplace for Learning, Innovation, and Growth*. Hoboken, NJ: John Wiley & Sons, 2019, p. XIV.

[84] re:Work. "Guide: Understand Team Effectiveness." https://rework.withgoogle.com/print/guides/5721312655835136.

[85] See https://www.adultdevelopmentstudy.org/publications.

[86] Robert Waldinger. "What Makes A Good Life? Lessons From the Longest Study on Happiness." TEDx Talk. November 2015. https://www.ted.com/talks/robert_waldinger_what_makes_a_good_life_lessons_from_the_longest_study_on_happiness.

[87] Office of the U.S. Surgeon General. "Our Epidemic of Loneliness and Isolation: The U.S. Surgeon General's Advisory on the Healing Effects of Social Connection and Community." 2023. https://www.hhs.gov/sites/default/files/surgeon-general-social-connection-advisory.pdf.

[88] Arthur C. Brooks. "Love Your Enemies: How Decent People Can Save America from the Culture of Contempt." Harvard Business School Faculty & Research. 2019. https://www.hbs.edu/faculty/Pages/item.aspx?num=56870.

[89] John M. Gottman and Nan Silver. *The Seven Principles for Making Marriage Work: A Practical Guide from the Country's Foremost Relationship Expert*. New York: Harmony Books, 2015.

[90] Clayton Christensen, James Allworth, and Karen Dillon. *How Will You Measure Your Life?* New York: HarperCollins, 2012.

[91] Isaac Newton. Letter to Robert Hooke. February 5, 1675.

ABOUT THE AUTHOR

Jeff Perry is a leadership and career expert known for helping individuals, teams, and organizations unlock their potential in all facets of life. Given his background in engineering, business, and leadership, he specializes in working with engineering and technical professionals, but the principles he shares are universal.

Jeff received a Bachelor of Science in Mechanical Engineering from Brigham Young University, and a Master of Business Administration from the University of Washington.

A sought-after teacher and speaker, Jeff is happily married to Robin. Together they have four children, and live in beautiful Pullman, Washington.

You can reach Jeff on LinkedIn at www.linkedin.com/in/jeffcperry or learn more at jeff-perry.com.

Made in the USA
Las Vegas, NV
30 January 2025